Application of Membrane Materials in Bioseparation and Downstream Processing

Application of Membrane Materials in Bioseparation and Downstream Processing

Guest Editors

Wei Zhang
Lingxue Kong

MDPI

Basel • Beijing • Wuhan • Barcelona • Belgrade • Novi Sad • Cluj • Manchester

Guest Editors
Wei Zhang
Bioprocessing Technology
Institute (BTI)
Agency for Science,
Technology and
Research (A*STAR)
Singapore
Singapore

Lingxue Kong
Institute for Frontier Materials
Deakin University
Geelong
Australia

Editorial Office
MDPI AG
Grosspeteranlage 5
4052 Basel, Switzerland

This is a reprint of the Special Issue, published open access by the journal *Membranes* (ISSN 2077-0375), freely accessible at: https://www.mdpi.com/journal/membranes/special_issues/9F92F5183G.

For citation purposes, cite each article independently as indicated on the article page online and as indicated below:

Lastname, A.A.; Lastname, B.B. Article Title. *Journal Name* **Year**, *Volume Number*, Page Range.

ISBN 978-3-7258-6772-1 (Hbk)
ISBN 978-3-7258-6773-8 (PDF)
https://doi.org/10.3390/books978-3-7258-6773-8

Contents

membranes

Editorial

Application of Membrane Materials in Bioseparation and Downstream Processing

Wei Zhang [1,*] and Lingxue Kong [2,*]

[1] Bioprocessing Technology Institute (BTI), Agency for Science, Technology and Research (A*STAR), 20 Biopolis Way, #06-01 Centros, Singapore 138668, Singapore
[2] Institute for Frontier Materials, Deakin University, Waurn Ponds, Geelong, VIC 3216, Australia
* Correspondence: zhang_wei@a-star.edu.sg (W.Z.); lingxue.kong@deakin.edu.au (L.K.)

1. Introduction

Membrane technologies have played an instrumental role in bioseparation and downstream processing. Membrane operations such as microfiltration, ultrafiltration, and virus filtration are key tools for clarification, concentration, buffer exchange, and viral clearance in biomanufacturing, as they offer mild, selective, and efficient separations for delicate biological products.

With advances in biotherapeutics development, membrane technologies have become increasingly pivotal in bioseparation and downstream processing. Compared with conventional packed-bed chromatography, a major advantage of membrane technologies is their convective mass transfer, which enables faster processing times and higher productivity. This advantage is particularly crucial for processing increasingly complex and fragile large biomolecules, such as bispecific antibodies, viruses, viral vectors, extracellular vesicles, and even cells.

In this Special Issue, Application of Membrane Materials in Bioseparation and Downstream Processing, published in Membranes, we have collected a diverse yet cohesive set of papers that demonstrate advances in the application of membrane technologies in bioseparation and downstream processing. These studies address highly relevant challenges ranging from conventional applications to emerging modalities, and from process optimization to novel membrane material design. Collectively, they underscore the pivotal role of membrane technologies in enabling efficient and robust separations in modern biomanufacturing.

2. Overview of Special Issue Contributions

To begin, several papers included in this Special Issue demonstrate advances in conventional applications of membrane technologies in bioseparation and downstream processing, including bioreactor clarification and virus filtration. Mostafavi et al. explore the potential to reduce membrane fouling by integrating a hydrocyclone as the primary clarification operation and investigate the impact of membrane morphology and pore size of integrated tangential flow filtration on overall clarification performance, offering insights that can improve throughput and robustness in early downstream processing stages. Isu et al. investigate the effectiveness of different membrane adsorbers in disrupting reversible aggregates that form during virus removal filtration and contribute to flux decline, thereby helping to guide the development of more effective virus filtration processes for monoclonal antibody production. Afzal et al. examine the effects of proteins on virus retention across different virus removal filters, providing important insights into the development and application of virus filtration in bioprocessing.

Secondly, this Special Issue also showcases novel membrane applications for more complex biological products, including bispecific antibodies, outer membrane vesicles, viral vectors, and cells. Zhao et al. evaluate the impact of the operating buffer system of loading samples on aggregate removal using hydrophobic interaction membrane chromatography, demonstrating that this technique operated in flow-through mode is an effective and robust approach for reducing aggregates during bispecific antibody purification. Li et al. develop a membrane filtration method to isolate *Escherichia coli* Nissle 1917–derived outer membrane vesicles of varying sizes, providing direct experimental evidence and deeper biological insights that outer membrane vesicles of different sizes exhibit heterogeneous properties. Sarmah and Husson demonstrate the feasibility of separating full and empty capsids of adeno-associated virus via ultrafiltration, directly addressing a long-standing bottleneck in viral vector downstream processing and offering a potential alternative solution. Schwarz et al. characterize membrane fouling during the separation of *Vibrio natriegens* biomass from culture broth using a cross-flow filtration plant with ceramic membranes, providing guidance for the design of economical filtration plants and strengthening the potential of establishing *V. natriegens* as a production host for large-scale industrial processes.

Lastly, other contributions included in this Special Issue also summarize recent breakthroughs in rational membrane design and shed light on future directions in the field. Boztepe et al. highlight the growing relevance of ion-exchange nanofibrous membranes in membrane chromatography for protein purification, underscore the importance of computational modelling as a viable predictive approach to guide membrane design and performance prediction and emphasize that integrating AI-guided design with high-throughput characterization could represent a future direction for membrane material innovation.

3. Conclusions

In summary, this Special Issue presents a diverse collection of papers on membrane materials in bioseparation and downstream processing, covering applications ranging from conventional bioreactor clarification and viral clearance to the purification of emerging modalities, including bispecific antibodies, outer membrane vesicles, viral vectors, and cells. Collectively, these studies demonstrate that membrane technologies have progressed from an auxiliary role to a central, enabling tool in biomanufacturing. We hope this Special Issue not only provides a timely update on the applications of membrane materials in bioseparation and downstream processing but also serves as a catalyst for continued innovation in membrane technologies.

Author Contributions: W.Z. and L.K. equally contributed to this Editorial. All authors have read and agreed to the published version of the manuscript.

Acknowledgments: We thank the editorial board of Membranes for their invaluable support.

Conflicts of Interest: The authors declare no conflicts of interest.

List of Contributions:

1. Zhao, P.; Qi, Y.; Gao, K. Removal of Aggregates During Bispecific Antibody Purification Using Hydrophobic Interaction Chromatography. *Membranes* **2025**, *15*, 299. https://doi.org/10.3390/membranes15100299.
2. Li, N.; Xin, H.; Deng, K. Separation and Characterization of Heterogeneity Among Various Sizes of Outer Membrane Vesicles Derived from the Probiotic *Escherichia coli* Nissle 1917. *Membranes* **2025**, *15*, 141. https://doi.org/10.3390/membranes15050141.

3. Schwarz, S.; Fan, R.; Ebrahimi, M.; Czermak, P. Efficient Separation of a Novel Microbial Chassis, *Vibrio natriegens*, from High-Salt Culture Broth Using Ceramic Ultrafiltration Membranes. *Membranes* **2025**, *15*, 121. https://doi.org/10.3390/membranes15040121.
4. Mostafavi, A.H.; Chu, L.-K.; Qian, X.; Smelko, J.P.; Zhang, D.; Zydney, A.; Wickramasinghe, S.R. Enhancing the Performance of Tangential Flow Microfiltration for Bioreactor Clarification. *Membranes* **2025**, *15*, 78. https://doi.org/10.3390/membranes15030078.
5. Isu, S.; Chen, S.-T.; Daneshpour, R.; Shirataki, H.; Strauss, D.; Zydney, A.L.; Qian, X.; Wickramasinghe, S.R. Enhancing Virus Filter Performance Through Pretreatment by Membrane Adsorbers. *Membranes* **2025**, *15*, 34. https://doi.org/10.3390/membranes15010034.
6. Sarmah, D.; Husson, S.M. A Novel Method for Separating Full and Empty Adeno-Associated Viral Capsids Using Ultrafiltration. *Membranes* **2024**, *14*, 194. https://doi.org/10.3390/membranes14090194.
7. Afzal, M.A.; Peles, J.; Zydney, A.L. Comparative Analysis of the Impact of Protein on Virus Retention for Different Virus Removal Filters. *Membranes* **2024**, *14*, 158. https://doi.org/10.3390/membranes14070158.
8. Boztepe, I.; Zhao, S.; Yang, X.; Kong, L. Toward Rational Design of Ion-Exchange Nanofiber Membranes: Meso-Scale Computational Approaches. *Membranes* **2026**, *16*, 5. https://doi.org/10.3390/membranes16010005.

membranes

MDPI

Article

Removal of Aggregates During Bispecific Antibody Purification Using Hydrophobic Interaction Chromatography

Puya Zhao *, Yue Qi * and Kai Gao

Shanghai AsymChem Biotechnology Co., Ltd., Building 8, No.12, Lane 855, Jinzheng Road, Jinshan Industrial Park, Shanghai 201507, China; gaokai@asymchem.com.cn
* Correspondence: zhaopuya0522@asymchem.com.cn (P.Z.); qiyue0982@asymchem.com.cn (Y.Q.)

Abstract

In the production of recombinant antibody/Fc-fusion proteins using mammalian cells, many aggregates often form alongside the target proteins, particularly with bispecific antibodies. To ensure the safety of biological products, it is essential to control the amount of aggregates within a specific range. A traditional downstream process typically involves using Protein A (ProA) resin to capture the target antibody, followed by two polishing steps to ensure purity; for instance, using an anion exchange chromatography (AEX) in flow-through mode and a cation exchange chromatography (CEX) in binding–elution mode. In this study, we choose a Dual Action Fab (DAF), which can bind two antigens and is prone to aggregation when expression in CHO (Chinese Hamster Ovary) cells. We introduce hydrophobic interaction membrane chromatography (HIMC) operating in flow-through mode, which enhances production efficiency while reducing costs and the risks associated with column packing. We evaluated the impact of the operating buffer system, as well as the pH and conductivity of the loading samples, on aggregate removal using HIMC. Additionally, we investigated the mechanism of aggregate binding and found that loading conditions had a limited impact on this process. Overall, our findings indicate that employing HIMC can achieve a 20% reduction in aggregate levels. These results demonstrate that HIMC in flow-through mode is an effective and robust approach for reducing aggregates during antibody purification.

Keywords: downstream process; membrane chromatography; quality by design; flow-through mode

1. Introduction

Bispecific antibodies (BsAbs), engineered to simultaneously engage two distinct epitopes on a single target or across different molecular entities, have emerged as transformative therapeutic modalities since their conceptual inception [1,2]. The evolutionary trajectory of BsAbs traces back to the 1960s, when pioneering studies demonstrated the feasibility of generating bispecific F(ab')2 molecules through re-association of antigen-binding fragments derived from heterologous polyclonal sera. A paradigm shift occurred with Köhler and Milstein's hybridoma breakthrough in 1975, enabling systematic exploration of rodent monoclonal antibody conjugation and hybridoma fusion technologies to achieve precise dual-targeting specificity [3]. Subsequent decades witnessed an exponential diversification of BsAb architectures, culminating in accelerated clinical translation. The recent surge in therapeutic BsAb development reflects their distinct pharmacological advantages over conventional monoclonal antibodies, including augmented target engagement precision, enhanced tumor microenvironment penetration, improved manufacturability profiles, and reduced immunogenic risks—attributes that position BsAbs as potent weapons in modern immuno-oncology [4,5]. Nevertheless, the inherent structural complexity of these molecules poses unique challenges in downstream processing, particularly regarding aggregate removal during purification—a critical quality attribute demanding innovative chromatographic optimization strategies.

Bispecific antibodies (BsAbs) in modern bio-therapeutic development are predominantly produced through recombinant DNA technologies, having evolved into multiple structural formats that broadly categorize into two classes: IgG-like constructs with functional Fc domains and non-IgG-like frameworks lacking Fc-mediated effector functions [1,2]. While IgG-like architectures dominate clinical pipelines due to their favorable pharmacokinetic profiles, their complex heterotetrametric structure—particularly in asymmetric variants requiring coordinated assembly of two distinct heavy chains (HCs) and light chains (LCs)—presents substantial manufacturing challenges [6,7]. During the expression of BsAbs, aggregates may form due to structural instability or miss-folding of amino acid sequences. Environmental factors such as temperature fluctuations, suboptimal pH conditions, and improper ionic strength during production or storage can further exacerbate aggregation phenomena [8,9]. These aggregates pose significant challenges to therapeutic applications: Aggregates may sterically hinder antigen-binding domains, diminishing target recognition efficacy and compromising therapeutic outcomes [10]. Particulate aggregates can enhance immunogenicity by activating dendritic cells and triggering anti-drug antibody responses, potentially leading to hypersensitivity reactions or reduced drug persistence [11]. Aggregation accelerates physical degradation pathways, jeopardizing product integrity during long-term storage and transportation [12,13].

The rapid advancement of BsAb therapeutics necessitates parallel innovations in downstream purification strategies to address unique product-related impurities that conventional mAb processes cannot effectively resolve. Traditional polishing approaches relying on ion exchange chromatography (IEX)–whether through anion exchange (AEX) flow-through (FT) or cation exchange (CEX) binding–elution(BE) modes–exploit isoelectric point (pI) differences between target molecules and impurities [14]. However, these methods face significant limitations in BsAb purification, particularly in resolving critical impurities like hole–hole homodimers that exhibit nearly identical pI values to the desired product. This challenge has driven the exploration of alternative separation mechanisms, with hydrophobic interaction chromatography (HIC) emerging as a promising orthogonal approach [15,16]. HIC leverages subtle differences in surface hydrophobicity between species, operating through either FT or BE modes, where protein–resin interactions are modulated by salt concentration gradients. While HIC has established utility in antibody purification, conventional implementations face inherent limitations, including constrained binding capacity, suboptimal throughput, and potential risks of protein denaturation due to strongly hydrophobic ligands—challenges that become particularly acute in industrial-scale BsAb manufacturing where both product quality and process efficiency are paramount [17]. These constraints underscore the need for innovative HIC optimization strategies to realize their full potential in next-generation BsAb purification platforms.

In pursuing next-generation purification technologies, researchers prioritize developing cost-effective and high-efficiency chromatography platforms. Notably, microporous membrane chromatography demonstrates transformative potential in addressing inherent limitations of conventional packed-bed systems, including excessive media/operational costs, diffusion-dominated intraparticle mass transfer mechanisms, and scalability challenges related to column packing consistency [18,19]. The structural innovations of membrane chromatography drive multi-level technical advancements: Modular configuration eliminates column packing processes and associated variability. Tortuosity-free flow paths reduce pressure drops by 2–3 orders of magnitude compared to packed beds. Convection-enhanced mass transfer enables 10–50× higher linear flow rates without compromising binding capacity, dramatically accelerating adsorption–desorption kinetics [19–23]. The Sartobind® Phenyl hydrophobic membrane adsorber, a pioneering innovation introduced, revolutionized biomolecule purification through its unique design [24]. The system integrates 30-layer radial flow capsules with an optimized 8 mm bed height, engineered to maximize operational efficiency. This breakthrough technology combines a specifically engineered macroporous membrane architecture with covalently immobilized phenyl ligands, creating a robust platform for high-throughput purification [25,26]. The membrane's structural design enables exceptional flow rate tolerance while maintaining high binding capacities, addressing critical limitations of traditional resin-based systems. Its hydrophilic stabilized regenerated cellulose matrix provides superior mechanical stability and chemical resistance, even under stringent cleaning conditions. Notably, the membrane demonstrates minimal nonspecific protein binding—a critical advantage when processing high-concentration biomolecules in lyotropic salt solutions. Consistent with HIC principles, the functionalized membrane exhibits a characteristic increase in binding capacity with rising ammonium sulfate concentrations. This predictable behavior ensures reliable process scalability and simplifies method development across diverse biomolecule purification applications.

In downstream biopharmaceutical purification processes, both BE and FT modes represent widely employed chromatographic techniques, though their applications differ significantly based on resin characteristics [27,28]. The FT mode is predominantly utilized in AEX to remove residuals like host cell protein (HCP) and host cell DNA (HCD) rather than aggregates. The BE mode finds principal application in HIC and CEX, where target molecule–resin interactions require controlled binding and subsequent elution to get rid of aggregates. However, it is well known that the BE mode of HIC has a lower dynamic binding capacity (DBC) and requires an extended processing time and various buffers [29]. Additionally, the HIC-BE mode necessitates substantial salt addition to elevate sample conductivity during loading, which may induce protein precipitation and/or denaturation, potentially compromising product integrity [30]. On the contrary, the FT mode has a substantially higher loading density, adsorbs only impurities, and requires no additional recovery equipment [14]. Therefore, in this study, we will explore the possibilities of HIC membranes for both BE and FT modes to improve the process efficiency as much as possible while ensuring the removal of impurities.

Given these considerations, combined Sartobind® phenyl membrane adsorber and FT mode, the efficiency can be greatly improved while removing the aggregates. This HIMC combines the advantages of HIC and membrane chromatography, ensuring no diffusional limitations, a shorter processing time, a high capacity, and excellent resolution [24]. When synergized with HIMC under convective flow regimes, this approach demonstrates transformative advantages. In this study, we evaluated the efficacy of CEX for aggregate removal, while comparatively investigating BE and FT modes in HIMC. The HIMC-FT mode demonstrated superior performance in aggregate clearance. Furthermore, a DOE approach was implemented to optimize loading conditions, capacity, recovery, and SEC purity, successfully establishing the optimal operational window.

2. Materials and Methods

2.1. Antibody and Its Properties

Chinese hamster ovary (CHO) glutamine synthetase knockout (GS−/−) host cell (HD-BIOP3, Horizon) were used to express recombinant antibodies with standard cell culture methods (stirred bioreactor) [31]. Cells were cultivated in batch mode: in 2 L Biostat® bioreactors (Sartorius Stedim Biotech, Göttingen, Germany) for 14 days. A Sarto-clear BT 1000 filter with a pore size of 0.22 μm (Sartorius Stedim Biotech) was used to filter the cells after undergoing two-step centrifugation. The antibody employed in this experiment was a BsAb and its properties are listed in Table 1.

Table 1. The antibody and its properties. The information includes the molecule class, pI and MW.

Molecule Class	pI	MW (kDa)
BsAb	8.15	150.52

Note: BsAb—bispecifc antibody, pI—Isoelectric point, MW—Molecular Weight.

2.2. Equipment

This experiment was carried out using the Cytiva ÄKTA Avant 150 chromatography system. The system comprises two positive displacement pumps (A and B) connected to the mixing chamber. The flow was routed via the mixing chamber to the column. The UNICORNTM (version 7.6) software (Cytiva, Uppsala, Sweden) was used to operate the pumps and collect and process the data.

2.3. Protein A Chromatography

MabSelect SuRe LX (Cytiva, Uppsala, Sweden) was used to capture antibodies from harvested cell culture fluid (HCCF). The loading capacity was 30.0 g/L in accordance with the volume of the resin. The retention time was set to 5 min, and samples for analysis ranged from the UV_{280} peak of 50 mAu/mm to the maximum and then to 50 mAu/mm. Chromatography was carried out with the buffers, retention time (RT), and procedure volume listed in Table 2.

Table 2. Procedure and buffers of protein A chromatography. There are experiment procedures, buffer and resistance time, and column volume for affinity chromatography.

Procedure	Buffer	RT (min)	Volume (CV)
Sanitization	1 M NaOH	5	3
Equilibration	50 mM Tris-HAc, 150 mM NaCl, pH 7.4		3
Elution	50 mM NaAc-HAc, pH 3.6		3
Regeneration	120 mM HAc		3

Note: CV—column volume.

2.4. AEX

AEX was carried out with POROS 50HQ resin (Thermo Fisher, Bedford, TX, USA). The loading capacity was 150 g/L. The retention time was set to 5 min. Chromatography was carried out with the buffers and procedures listed in Table 3.

Table 3. Procedure and Buffers of AEX. There are experiment procedures, buffer and resistance time, and column volume for AEX.

Procedure	Buffer	RT (min)	Volume (CV)
Sanitization	1 M NaOH	5	3
Equilibration	50 mM NaAc-HAc, pH 5.5		3
Strip	50 mM NaAc-HAc, 500 mM NaCl, pH 5.5		3

Note: CV—column volume.

2.5. CEX

CEX was carried out with Capto S ImpAct resin (Cytiva, Uppsala, Sweden). The loading capacity was 40 g/L. The retention time was set to 5 min. The elution buffer was a

0–100% gradient combination of A: 50 mM NaAc-HAc, pH5.5 and B: 50 mM NaAc-HAc, 500 mM NaCl, pH5.5. Chromatography was carried out with the buffers and procedures listed in Table 4.

Table 4. Procedure and buffers of CEX. There are experiment procedures, buffer and resistance time, and column volume for CEX.

Procedure	Buffer	RT (min)	Volume (CV)
Sanitization	1 M NaOH		3
Equilibration	50 mM NaAc-HAc, pH 5.5		3
Elution	A:50 mM NaAc-HAc, pH 5.5 B:50 mM NaAc-HAc, 500 mM NaCl, pH 5.5	5	0–100%B, 20 CV
Regeneration	50 mM NaAc-HAc, 500 mM NaCl, pH 5.5		3

Note: CV—column volume.

2.6. HIMC

Sartobind® Phenyl (Sartorius Stedim, Göttingen, Germany) was a membrane used in the flow-through mode of the Hydrophobic Interaction Membrane Chromatography (HIMC), which has a volume of 3 mL, 8 mm bed height. Prior to the experiment, the loading sample's conductivity and pH were adjusted in accordance with the equilibration buffer. The resin's loading capability was 15 g/L. The fraction was collected between 50 mAu/mm and 100 mAu/mm, and the flow-through product was examined using SEC-HPLC. The BE mode (binding–elution mode) was also applied for evaluation purposes, chromatography was carried out with the buffers and procedures listed in Table 5. And the FT mode chromatography was carried out with the buffers and procedures listed in Table 6.

Table 5. Procedure and buffers of HIMC (BE mode). There are experiment procedures, buffer and resistance time, and column volume for HIC.

Procedure	Buffer	RT (min)	Volume (MV)
Sanitization	1 M NaOH		15
Equilibration	50 mM NaAc-HAc, 900 mM $(NH_4)_2SO_4$, pH 5.5		15
Elution	A: 50 mM NaAc-HAc, 1.05 M $(NH_4)_2SO_4$, pH 5.5 B: 50 mM NaAc-HAc, pH 5.5	0.3	0–100%B, 20 MV
Regeneration	HPW		15

Note: MV—membrane volume, HPW—high purity water.

Table 6. Procedure and buffers of HIMC (FT mode). There are experiment procedures, buffer and resistance time, and column volume for HIC.

Procedure	Buffer	RT (min)	Volume (MV)
Sanitization	1 M NaOH		15
Equilibration	50 mM NaAc-HAc, 450–900 mM$(NH_4)_2SO_4$, pH 5.5	0.3	15
Regeneration	HPW		15

Note: MV—membrane volume, HPW—high purity water.

2.7. Analytical Techniques

2.7.1. Determination of DBC

The binding capacity studies were carried out as follows: the membranes were initially cleansed with equilibration buffer for 15 membrane volumes (MVs) at the RT of 18 s. The HIMC Load (pH 5.5, 140 mS/cm, the sample conductivity was adjusted by 3 M $(NH_4)_2SO_4$ solution) was then continuously loaded on Sartobind® Phenyl with a membrane volume of 3 mL at a residence time of 18 s until the UV absorbance at 280 nm showed a 10% breakthrough. The concentration of the flow-through samples was measured using UV Spectrophotometer-NanoDrop one (Thermo Fisher, Delaware, DE, USA) with the molecule-specific extinction coefficients. No dilution was required for any of the samples. After that, the membrane was washed with 15 MVs of equilibration buffer, and the proteins were flushed out with 15 MVs of elution buffer. After stripping, the membrane was sanitized with 15 MVs of sodium hydroxide, neutralized with HPW, and stored in 20% v/v ethanol.

The amount of protein loaded into the membrane was calculated from the following equation:

$$\text{Protein binding capacity} = C_0 \times (V_L - V_0)/V_M, \quad (1)$$

Note: C_0—protein concentration in the sample (mg/mL), V_L—accumulated volume per fraction (mL), V_0—system void volume (mL), V_M—membrane volume (mL).

2.7.2. Design of Experiment (DoE) Screening Study Design and Data Analysis

The conductivity and pH of the applied sample, along with loading capacity, were the three factors at two levels in the multivariate analysis. A complete factorial design was used to examine these three factors, requiring 8 experiments. Additionally, 3 center point experiments were added to improve accurate estimation of experimental errors and ensure that there are (sufficient) degrees of freedom to analyze experimental errors [32]. Since the complete factorial design allows for all interactions and linear effects, each factor may be assessed independently. The experimental setup is displayed in Figure 1 and Table 7. Optimal factor settings for the process were obtained by entering the findings of each experimental point into JMP® (18) software after the optimization trial series was finished. Statistical examination of the generated data was part of the review process.

Table 7. Experimental study design. There are three factors: pH, conductivity and capacity.

Mode	pH	Conductivity (mS/cm)	Capacity (g/L)
0 0 0	6.5	90	40
0 0 0	6.5	90	40
− + −	5.5	120	20
− − +	5.5	60	60
+ − −	7.5	60	20
− − −	5.5	60	20
+ + −	7.5	120	20
+ + +	7.5	120	60
0 0 0	6.5	90	40
+ − +	7.5	60	60
− + +	5.5	120	60

Note: "0" represents the center point; "+" represents the high level; "−" represents the low level.

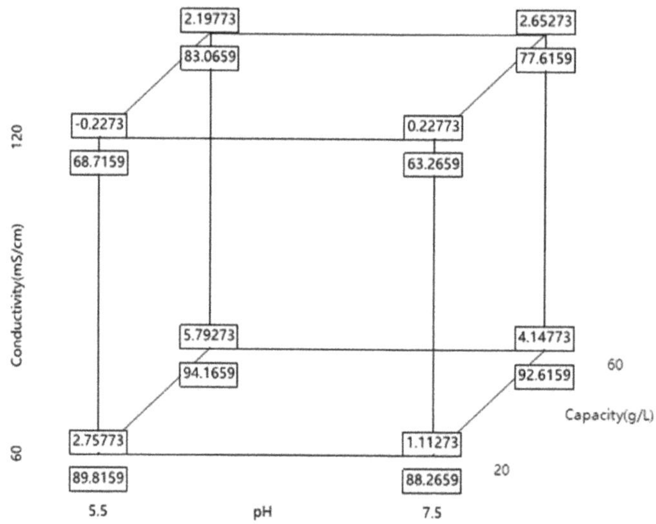

Figure 1. Schematic representation of the conducted design of experiments for the conductivity, pH, and capacity.

2.7.3. Recovery

Spectrophotometric analysis was used to determine the BsAb concentration. Product recovery was calculated from Equation (2):

$$\text{Recovery} = \text{mass of BsAb(monomers) in elution} / \text{mass of BsAb(monomers) applied to column or membrane} \quad (2)$$

2.7.4. Size Exclusion Chromatography-High-Performance Liquid Chromatography (SEC-HPLC)

Agilent 1100 Series was applied as the SEC-HPLC equipment, and absorbance at 280 nm was used to quantify the amounts of monomers and aggregates. Samples were injected using 200 mM potassium phosphate, which contains 250 mM potassium chloride (pH 6.8) as the running buffer into a SEC column (250 Å, 5 μm, 7.8 mm × 300 mm, Waters, Taunton, MA, USA). The working flow rate was set to 0.8 mL/min. By combining the peak regions of the early-eluting aggregate peak(s), late-eluting fragment peak(s), and monomer peak, the ratios of monomers and aggregates were calculated, which indicated the aggregate levels of the product.

3. Results and Discussion

3.1. The HMWs Removal Ability Evaluation of the CEX Column

CEX and HIC are common methods to separate monomers and aggregates in antibody industry production. Compared with HIC, CEX has a lower requirement for loading conductivity and higher capacity [29,33]. Thus, the CEX (Capto S Impact, Cytiva) column was primarily tested to separate aggregates and monomers. The loading condition of CEX is pH5.5, conductivity 4.0 mS/cm, and the capacity is 30.0 g/L. Figure 2 shows the CEX profile in the elution phase. Based on Figure 2, the fraction eluate was CV 4–9. However, CV 6–8 contained HMW levels over 10.0%. Figure 2b shows a shoulder peak in the fraction eluate CV7. SEC-HPLC results of CEX load and fraction 4–7 collection are shown in Table 8. Thus, the resolution between aggregates and monomers was poor. Combined with Figure 2 and Table 8, it was challenging to separate HMWs and monomers based on the difference in pI between aggregates and monomers by the change in elution conductivity in CEX. Thus, HIC would be tested based on the hydrophobicity difference between monomers and aggregates.

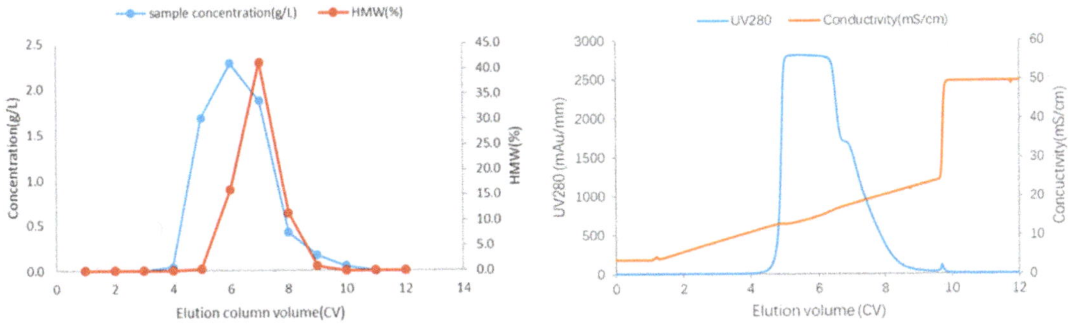

(a) (b)

Figure 2. (**a**) CEX elution profile; (**b**) CEX UV280 adsorption and conductivity chromatography in the elution phase. Both showed a poor resolution between aggregates and monomers.

Table 8. The quality data of CEX. The recovery and SEC-HPLC results of CEX load and collection eluate.

Sample	Recovery (%)	SEC-HPLC			CE-NR		HCP (ppm)	HCD (ppb)
		HMWs (%)	Monomers (%)	LMWs (%)	Purity (%)	LMWs (%)		
CEX load	N/A	20.9	79.1	ND	96.1	3.9	36.9	<0.3
CEX eluate 4–7	90.1	18.7	81.3	ND	96.2	3.8	3.6	<0.2

3.2. *The Evaluation of FT and BE Mode of HIMC*

Compared with column chromatography, membrane chromatography has an advantage of short residence time, which can decrease the single experiment run time and increase research efficiency [21]. Thus, HIMC (Sartobind® Phenyl, 3 mL) was attempted to remove aggregates in this study. To evaluate the aggregates removal ability of HIMC fully, both BE and FT modes were investigated.

3.2.1. The Dynamic Binding Capacity of HIMC for BE Mode

For HIMC BE mode, dynamic binding capacity (DBC) study was need to evaluate to find out the maximum loading density of the HIMC. To increase the binding capacity between sample and the ligand of HIMC, the loading conductivity should be large enough to increase the sample hydrophobicity. The loading condition of the HIMC was pH 5.50 and 140 mS/cm in this study. Figure 3 shows the DBC determination breakthrough curve at the RT 0.3 min. When the breakthrough was 10%, the corresponding DBC of HIMC was 18.2 g/L membrane. The final capacity of HIMC is $DBC_{10\%}$ × safety factor (0.8), thus, the capacity of HIMC in BE mode is 18.2 g/L × 0.8 = 14.5 g/L.

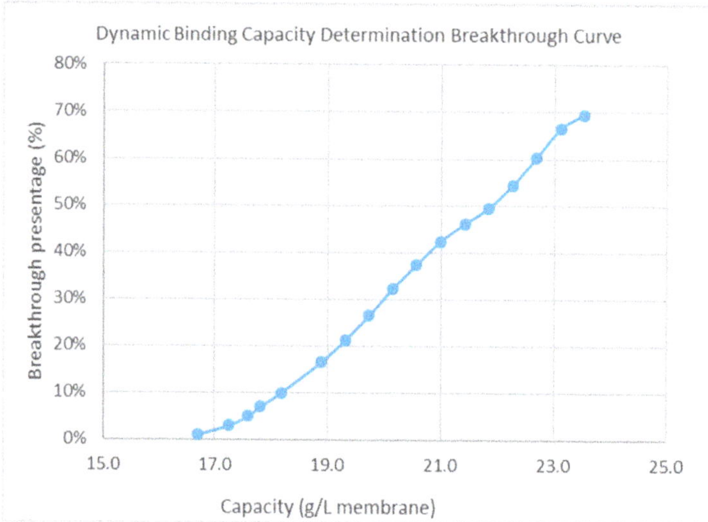

Figure 3. The dynamic binding capacity determination breakthrough curve of HIMC.

3.2.2. The Aggregate Removal Ability of HIMC for BE Mode and FT Mode

For the HIMC BE mode test, the loading density was 14.5 g/L, and the loading condition was pH5.5, 140 mS/cm. Based on previous research [34,35], the FT mode for column purification can tolerate a larger capacity than the BE mode, so for the initial attempt for the HIMC FT mode, the test conditions were the capacity of 30.0 g/L, and the loading condition was pH5.5, 90 mS/cm. The capacity and loading conditions are summarized in Table 9. Figure 4 shows the protein concentration, HMW level, and chromatography of HIMC BE mode in the Elution phase. Table 10 shows the recovery and SEC-HPLC purity for HIMC load, HIMC eluate, and FT pool. As shown in Tables 8 and 10, although the capacity of HIMC BE mode is nearly half of the CEX BE mode and the recovery is less than 15.9%, the SEC monomers purity of HIMC BE mode is higher than that of CEX by about 14.6%. Compared with Figure 2, Figure 4 shows that HIMC BE mode has a higher resolution between aggregates and monomers than CEX BE mode. It shows that separation between aggregates and monomers by hydrophobicity is a potential method when the pI difference between monomers and aggregates is similar. The work principle of BE mode and FT mode of HIMC is shown in Figure 5. As shown in Table 10, HIMC FT mode has nearly twice its capacity, less 1.1% HMWs, and a higher 6.0% recovery than HIMC BE mode. It shows that HIMC FT mode has a higher potential than HIMC BE mode in the recovery, capacity, and aggregates removal ability section. HIMC FT mode optimization will be discussed in Section 3.3.

Table 9. The summary of capacity and loading condition of HIMC BE and FT mode.

Mode	Capacity (g/L Membrane Volume)	Loading pH/Conductivity (mS/cm)
Binding–elution	14.5	5.5/140.0
Flow-through	30.0	5.5/70.0

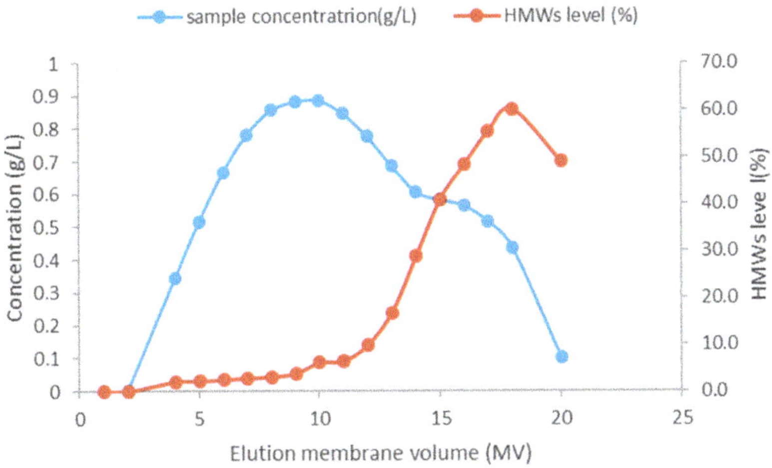

Figure 4. HIMC of BE mode elution profile.

Table 10. The quality data of HIMC. The recovery and SEC-HPLC results of HIMC load, collection eluate MV 4–14 in BE mode and collection FT in FT mode.

Sample	Recovery (%)	SEC-HPLC			CE-NR		HCP (ppm)	HCD (ppb)
		HMWs (%)	Monomers (%)	LMWs (%)	Purity (%)	LMWs (%)		
HIMC load	N/A	20.9	79.1	ND	96.1	3.9	36.9	<0.3
HIMC eluate 4–17	74.2	4.1	95.9	ND	96.1	3.9	3.1	<0.3
HIMC flow-through	80.2	3.0	97.0	ND	96.3	3.7	3.4	<0.3

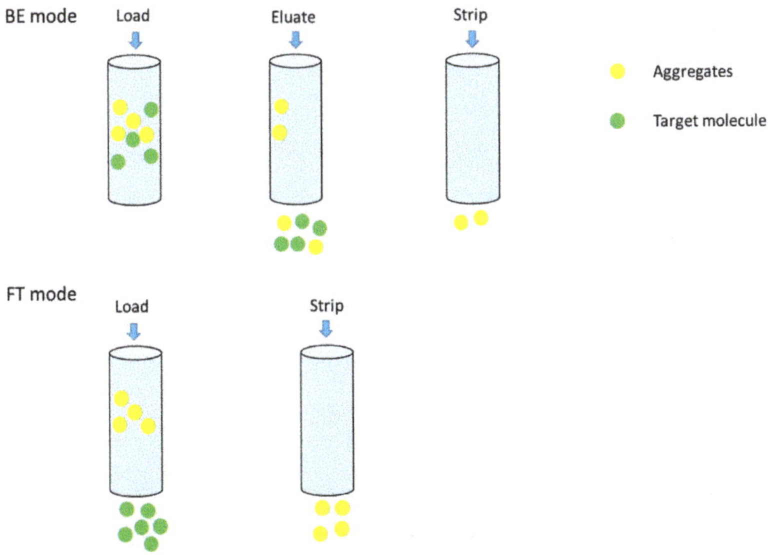

Figure 5. HIMC BE and FT mode working principle.

3.3. *The Optimization of HIMC FT Mode by DoE Study and Analysis*

3.3.1. Experimental Strategy, Results, and Statistical Analysis of DoE

In the HIMC FT mode, aggregates can be attached to HIMC ligands by hydrophobicity interaction and monomers flow through directly. The loading density and condition, including pH and conductivity, may affect the attachment between aggregates and HIMC ligands. Thus, the DoE study was utilized to optimize the loading condition and loading density parameters. The DoE recovery and SEC-HPLC results are summarized in Table 11.

Table 11. The quality data of DOE. The recovery and SEC-HPLC results of DoE.

Mode	pH	Conductivity (mS/cm)	Capacity (g/L)	HMW (%)	Recovery (%)
0 0 0	6.5	90	40	1.81	83.5
0 0 0	6.5	90	40	1.79	83.1
− + −	5.5	120	20	0.31	67.3
− − +	5.5	60	60	6.16	94.1
+ − −	7.5	60	20	1.48	88.2
− − −	5.5	60	20	2.80	89.5
+ + −	7.5	120	20	0.10	64.3
+ + +	7.5	120	60	3.19	76.2
0 0 0	6.5	90	40	1.76	81.5
+ − +	7.5	60	60	4.19	92.3
− + +	5.5	120	60	2.07	84.1

Note: "0" represents the center point; "+" represents the high level; "−" represents the low level.

Statistical analysis was conducted on the responses of HMW level and recovery. JMP® (18) software was applied for statistical analysis and a least squares fitting algorithm was used to obtain a good model for delta pressure of the HMW level (R^2 = 0.94, model p-value = 0.0044) and recovery (R^2 = 0.99, model p-value < 0.0001). The results of statistical analysis were shown in Figure 6. Further evaluation of the varied process parameter effects and interactions would be discussed in Sections 3.3.3 and 3.3.4.

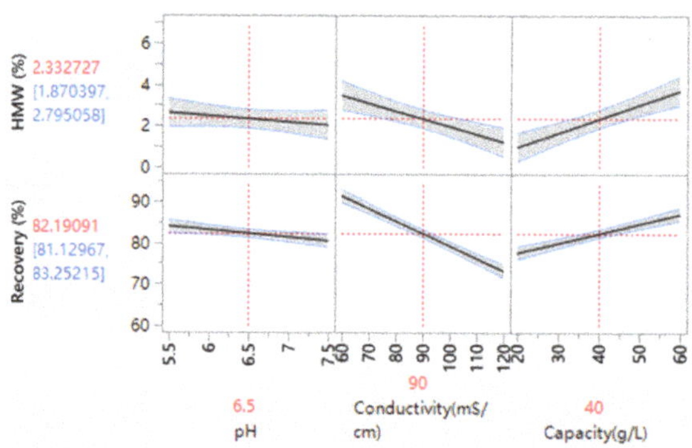

Figure 6. Prediction profiler depiction of changes in the HMW level and recovery as a function of the factors studied in the DoE. Note: vertical dotted lines represent the current value or current setting of each independent variable (X-variable); horizontal dotted lines indicate the current predicted value of each dependent variable (Y-variable) based on the current settings of the independent variables; the red fonts indicate the current value; the blue fonts represent the confidence intervals.

3.3.2. The Load pH and Conductivity for HIMC FT Mode

As shown in Figure 5, the pH has a slight influence on HMW levels and recovery. However, the loading conductivity has a significant effect on HMW levels and recovery.

With the loading conductivity increasing, the recovery will decrease, and the aggregates will also decrease in the HIMC FT pool. The reason may be that the high loading conductivity increases monomer and aggregates' hydrophobicity interactions with HIMC ligands. With the strong interaction between aggregates and HIMC ligands, the HMW level will decrease in the FT pool. However, parts of monomers were attached with HIMC ligands, as well. As for loading pH, the increase in pH will slightly increase the monomers and aggregates' hydrophobicity, especially if the loading conductivity is low (60 mS/cm), as shown in Table 11. If the loading conductivity increases, the pH variance influence of hydrophobic loading samples is weak. Thus, the load conductivity significantly influences HMW removal and recovery of HIMC FT mode more than the loading pH.

3.3.3. The Capacity for HIMC FT Mode

The loading density also plays an essential role in HMW removal and recovery for HIMC FT mode, as shown in Figure 5. If the capacity increases, the aggregate removal ability will decrease, and the recovery of HIMC FT mode will increase. The reason may be that the binding sites between aggregates and HIMC ligands are in a specific range as the ligands of HIMC are limited. If the hydrophobicity of loading samples increases, the interaction between aggregates and HIMC ligands increases, and more aggregates can be attached to the same amount of HIMC ligands. Thus, increasing loading conductivity will increase the HIMC FT capacity without affecting recovery and HMW removal ability. Also, if the loading conductivity is so high, proteins may precipitate, resulting in protein loss. It is necessary for the protein to test conductivity adjuster tolerance before beginning the HIMC FT study.

3.3.4. The Design Space of HIMC FT Mode

Based on the balance of recovery and HMW removal ability, the HMW level of less than 3.0% and recovery over 80.0% were set as the response requirements for HIMC FT mode. The design space was calculated by contour and surface plotters, as shown in Figure 7. Combined, the consideration of loading density and robustness of the process for scale-up, capacity between 37.0 and 45.0 g/L, loading pH 6.5, and loading conductivity between 78.0 and 91.0 mS/cm were recommended as the final process parameters.

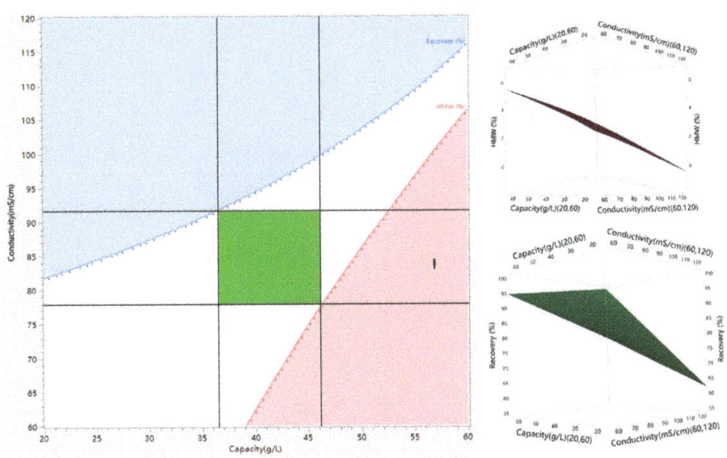

Figure 7. Design space based on the HMW level of less than 3.0% and recovery over 80.0%. The green area represents design space, the red area represents an HMW level of over 3.0%, and the blue area represents less than 80.0% recovery.

4. Conclusions

In conclusion, this study investigates the aggregate removal ability for the CEX column and novel HIMC in both BE and FT modes. Compared with the traditional polishing step, CEX column purification, HIMC shows significant potential for separating aggregates from monomers, especially for complicated molecules, like BsAb. Furthermore, this study explores an innovative approach to the use of HIMC, the FT mode, which compensates for HIMC's low load capacity and low resolution due to its own structural design. Compared to the HIMC BE mode, the FT mode has a higher aggregate removal ability. Meanwhile, the recovery and capacity of the HIMC FT mode are higher than that of BE mode. By HIMC FT mode, the HMW level can be decreased from 21.9% to less than 3.0% with a recovery of over 80.0%, which proves that HIMC FT mode is a robust and effective method for antibody purification production. This study only describes the removal of aggregates from a symmetric bsAb. Future research can focus on developing methods to remove aggregates from different types of molecules.

Author Contributions: Conceptualization, P.Z. and Y.Q.; formal Analysis, Y.Q.; writing—original draft preparation, Y.Q.; writing—review & editing, P.Z., Y.Q., and K.G. All authors have read and agreed to the published version of the manuscript.

Funding: This research received no external funding.

Institutional Review Board Statement: Not applicable.

Data Availability Statement: The original contributions presented in this study are included in the article. Further inquiries can be directed to the corresponding authors.

Conflicts of Interest: The authors, Puya Zhao, Yue Qi, and Kai Gao, are employees of Shanghai AsymChem Biotechnology Co., Ltd. The paper reflects the views of the scientists and not the company.

Abbreviations

The following abbreviations are used in this manuscript:

BsAbs	Bispecific antibodies
ProA	Protein A
DAF	Dual Action Fab
CHO	Chinese Hamster Ovary
AEX	Anion exchange chromatography
CEX	Cation exchange chromatography
HIMC	Hydrophobic interaction membrane chromatography
HCs	Heavy chains
LCs	Light chains
IEX	Ion exchange chromatography
PI	Isoelectric point
HIC	Hydrophobic interaction chromatography
mAbs	Monoclonal antibodies
MW	Molecular weight
BT	Binding–elution
FT	Flow-through
CV	Column volume
MV	Membrane volume
HPW	High-purity water
HMWs	High molecular weights
C_0	Protein concentration in the sample (mg/mL)
V_L	Accumulated volume per fraction (mL)
V_0	System void volume (mL)

V_M	Membrane volume (mL)
DoE	Design of experiment
SEC	Size exclusion chromatography
HPLC	High-performance liquid chromatography
HCD	Host cell DNA
HCP	Host cell protein
RT	Retention time

References

1. Guo, G.; Han, J.; Wang, Y.; Li, Y. A potential downstream platform approach for WuXiBody-based IgG-like bispecific antibodies. *Protein Expr. Purif.* **2020**, *173*, 105647. [CrossRef] [PubMed]
2. Li, Y.; Wang, Y.; Shen, P.; Zhou, W. A roadmap for IgG-like bispecific antibody purification. In *Approaches to the Purification, Analysis and Characterization of Antibody-Based Therapeutics*; Elsevier: Amsterdam, The Netherlands, 2020; pp. 167–179.
3. Nie, S.; Wang, Z.; Moscoso-Castro, M.; D'Souza, P.; Lei, C.; Xu, J.; Gu, J. Biology drives the discovery of bispecific antibodies as innovative therapeutics. *Antib. Ther.* **2020**, *3*, 18–62. [CrossRef]
4. Herrera, M.; Pretelli, G.; Desai, J.; Garralda, E.; Siu, L.L.; Steiner, T.M.; Au, L. Bispecific antibodies: Advancing precision oncology. *Trends Cancer* **2024**, *10*, 893–919. [CrossRef]
5. Rader, C. Bispecific antibodies in cancer immunotherapy. *Curr. Opin. Biotechnol.* **2020**, *65*, 9–16. [CrossRef] [PubMed]
6. Andrade, C.; Arnold, L.; Motabar, D.; Aspelund, M.; Tang, A.; Hunter, A.; Chung, W.K. An Integrated Approach to Aggregate Control for Therapeutic Bispecific Antibodies Using an Improved Three Column Mab Platform-Like Purification Process. *Biotechnol. Prog.* **2018**, *35*, e2720. [CrossRef]
7. Chen, T.; Han, J.; Guo, G.; Wang, Q.; Wang, Y.; Li, Y. Monitoring removal of hole-hole homodimer by analytical hydrophobic interaction chromatography in purifying a bispecific antibody. *Protein Expr. Purif.* **2019**, *164*, 105457. [CrossRef]
8. Li, W.; Prabakaran, P.; Chen, W.; Zhu, Z.; Feng, Y.; Dimitrov, D.S. Antibody Aggregation: Insights from Sequence and Structure. *Antibodies* **2016**, *5*, 19. [CrossRef]
9. Prabakaran, R.; Rawat, P.; Thangakani, A.M.; Kumar, S.; Gromiha, M.M. Protein aggregation: In silico algorithms and applications. *Biophys. Rev.* **2021**, *13*, 71–89. [CrossRef]
10. Hassan, M.; Shahzadi, S.; Li, M.S.; Kloczkowski, A. Prediction and Evaluation of Protein Aggregation with Computational Methods. *Methods Mol. Biol.* **2025**, *2867*, 299–314. [PubMed]
11. Chandel, T.I.; Zaman, M.; Khan, M.V.; Ali, M.; Rabbani, G.; Ishtikhar, M.; Khan, R.H. A mechanistic insight into protein-ligand interaction, folding, misfolding, aggregation and inhibition of protein aggregates: An overview. *Int. J. Biol. Macromol.* **2018**, *106*, 1115–1129. [CrossRef]
12. Eftekharzadeh, B.; Hyman, B.T.; Wegmann, S. Structural studies on the mechanism of protein aggregation in age related neurodegenerative diseases. *Mech. Ageing Dev.* **2016**, *156*, 1–13. [CrossRef]
13. Sontag, E.M.; Samant, R.S.; Frydman, J. Mechanisms and Functions of Spatial Protein Quality Control. *Annu. Rev. Biochem.* **2017**, *86*, 97–122. [CrossRef] [PubMed]
14. Chen, S.W.; Hoi, K.M.; Mahfut, F.B.; Yang, Y.; Zhang, W. Effective flow-through polishing strategies for knob-into-hole bispecific antibodies. *Bioresour. Bioprocess.* **2022**, *9*, 98. [CrossRef]
15. Hall, T.; Kelly, G.M.; Emery, W.R. Use of mobile phase additives for the elution of bispecific and monoclonal antibodies from phenyl based hydrophobic interaction chromatography resins. *J. Chromatogr. B* **2018**, *1096*, 20–30. [CrossRef]
16. Schmitz, F.; Kruse, T.; Minceva, M.; Kampmann, M. Integrated double flow-through purification of monoclonal antibodies using membrane adsorbers and single-pass tangential flow filtration. *Biochem. Eng. J.* **2023**, *195*, 108913. [CrossRef]
17. Ghosh, R.; Wang, L. Purification of humanized monoclonal antibody by hydrophobic interaction membrane chromatography. *J. Chromatogr. A* **2006**, *1107*, 104–109. [CrossRef]
18. Osuofa, J.; Husson, S.M. Comparative Evaluation of Commercial Protein A Membranes for the Rapid Purification of Antibodies. *Membranes* **2023**, *13*, 511. [CrossRef] [PubMed]
19. Grünberg, M.; Kuchemüller, K.B.; Töppner, K.; Busse, R.A. Scalable, Robust and Highly Productive Novel Convecdiff Membrane Platform for mAb Capture. *Membranes* **2022**, *12*, 677. [CrossRef] [PubMed]
20. Steegmüller, T.; Kratky, T.; Gollwitzer, L.; Schwaminger, S.P.; Berensmeier, S. Development of a New Affinity Gold Polymer Membrane with Immobilized Protein A. *Membranes* **2024**, *14*, 31. [CrossRef]
21. Yang, S.; Braczkowski, R.; Chen, S.-H.; Busse, R.; Li, Y.; Fabri, L.; Bekard, I.B. Scalability of Sartobind® Rapid A Membrane for High Productivity Monoclonal Antibody Capture. *Membranes* **2023**, *13*, 815. [CrossRef]

22. Chen, J.; Yu, B.; Cong, H.; Shen, Y. Recent development and application of membrane chromatography. *Anal. Bioanal. Chem.* **2022**, *415*, 45–65. [CrossRef]
23. Boi, C.; Malavasi, A.; Carbonell, R.G.; Gilleskie, G. A direct comparison between membrane adsorber and packed column chromatography performance. *J. Chromatogr. A* **2020**, *1612*, 460629. [CrossRef] [PubMed]
24. Fraud, N.; Faber, R.; Kiss, C.; Demmer, W.; Hoerl, H.H.; Fischer-Fruehholz, S. Hydrophobic-interaction membrane chromatography for large-scale purification of biopharmaceuticals. *BioProcess Int.* **2009**, *7*, 30–35. Available online: https://www.bioprocessintl.com/chromatography/hydrophobic-interaction-membrane-chromatography-for-large-scale-purification-of-biopharmaceuticals (accessed on 1 June 2009).
25. Wang, Y.; Bhaskar, U.; Chennamsetty, N.; Noyes, S.; Guo, J.; Song, Y.; Lewandowski, A.; Ghose, S. Hydrophobic interaction chromatography in continuous flow-through mode for product-related variant removal. *J. Chromatogr. A* **2024**, *1736*, 465356. [CrossRef]
26. Qu, Y.; Bekard, I.; Hunt, B.; Black, J.; Fabri, L.; Gras, S.L.; Kentish, S.E. The Transition from Resin Chromatography to Membrane Adsorbers for Protein Separations at Industrial Scale. *Sep. Purif. Rev.* **2023**, *53*, 351–371. [CrossRef]
27. Lorek, J.K.; Karkov, H.S.; Matthiesen, F.; Dainiak, M. High throughput screening for rapid and reliable prediction of monovalent antibody binding behavior in flowthrough mode. *Biotechnol. Bioeng.* **2024**, *121*, 2332–2346. [CrossRef]
28. Chen, S.T.; Wickramasinghe, S.R.; Qian, X. Electrospun Weak Anion-exchange Fibrous Membranes for Protein Purification. *Membranes* **2020**, *10*, 39. [CrossRef]
29. Fekete, S.; Veuthey, J.-L.; Beck, A.; Guillarme, D. Hydrophobic interaction chromatography for the characterization of monoclonal antibodies and related products. *J. Pharm. Biomed. Anal.* **2016**, *130*, 3–18. [CrossRef]
30. Jungbauer, A.; Machold, C.; Hahn, R. Hydrophobic interaction chromatography of proteins. *J. Chromatogr. A* **2005**, *1079*, 221–228. [CrossRef]
31. Reinhart, D.; Damjanovic, L.; Kaisermayer, C.; Sommeregger, W.; Gili, A.; Gasselhuber, B.; Castan, A.; Mayrhofer, P.; Grünwald-Gruber, C.; Kunert, R. Bioprocessing of Recombinant CHO-K1, CHO-DG44, and CHO-S: CHO Expression Hosts Favor Either mAb Production or Biomass Synthesis. *Biotechnol. J.* **2019**, *14*, e1700686. [CrossRef] [PubMed]
32. Rüttimann, B.; Wegener, K. The Power of DOE: How to Increase Experimental Design Success and Avoid Pitfalls. *J. Serv. Sci. Manag.* **2015**, *8*, 250–258. [CrossRef]
33. Marek, W.; Muca, R.; Woś, S.; Piątkowski, W.; Antos, D. Isolation of monoclonal antibody from a Chinese hamster ovary supernatant. II: Dynamics of the integrated separation on ion exchange and hydrophobic interaction chromatography media. *J. Chromatogr. A* **2013**, *1305*, 64–75. [CrossRef] [PubMed]
34. Ghose, S.; Tao, Y.; Conley, L.; Cecchini, D. Purification of monoclonal antibodies by hydrophobic interaction chromatography under no-salt conditions. *mAbs* **2013**, *5*, 795–800. [CrossRef] [PubMed]
35. Chen, J.; Tetrault, J.; Ley, A. Comparison of standard and new generation hydrophobic interaction chromatography resins in the monoclonal antibody purification process. *J. Chromatogr. A* **2008**, *1177*, 272–281. [CrossRef]

membranes

MDPI

Article

Separation and Characterization of Heterogeneity Among Various Sizes of Outer Membrane Vesicles Derived from the Probiotic *Escherichia coli* Nissle 1917

Ning Li *, Hongbo Xin and Keyu Deng *

The National Engineering Research Center for Bioengineering Drugs and the Technologies, Jiangxi Province Key Laboratory of Bioengineering Drugs, Institute of Translational Medicine, Jiangxi Medical College, Nanchang University, Nanchang 330031, China; xinhb@ncu.edu.cn

* Correspondence: lining@ncu.edu.cn (N.L.); dky@ncu.edu.cn (K.D.)

Abstract: Outer membrane vesicles (OMVs) are extracellular vesicles secreted by Gram-negative bacteria with diameters of 20–250 nm. OMVs contain various biologically active substances from their parent bacteria, such as proteins, lipids, and nucleic acids. *Escherichia coli* Nissle 1917 (EcN) is a Gram-negative probiotic that resides in the human intestine. EcN-derived OMVs are pivotal in modulating intestinal immune responses. However, few studies have addressed the heterogeneity of EcN-derived OMVs in terms of size, significantly limiting the research on their clinical applications. Currently, there are a lack of feasible methods for obtaining EcN-derived OMVs of different sizes. To address this knowledge gap, we developed a membrane filtration method to isolate EcN-derived OMVs of varying sizes. In this study, we first used gradient filtration to isolate high-purity EcN-derived OMVs and conducted a proteomic analysis. Subsequently, we used membrane filtration to separate the EcN-derived OMVs by size. We successfully obtained EcN-derived OMVs of three specific sizes: <50 nm, 50–100 nm, and 100–300 nm. We then performed proteomic analyses of these EcN-derived OMVs and compared their protein profiles. Finally, we compared the ability of each EcN-derived OMV type to induce RAW264.7 macrophages to secrete the pro-inflammatory factor interleukin (IL)-1β and the anti-inflammatory factor IL-10. The EcN-derived OMVs contained 646 different proteins overall; those of different sizes contained different protein types. Among them, the EcN-derived OMVs in the <50 nm group contained significantly fewer proteins (262 different types in total) than those in the 50–100 nm (1603 types) and 100–300 nm (1568 types) groups. Furthermore, the <50 nm group had fewer membrane proteins (40) than the 50–100 nm (215) and 100–300 nm (209) groups. We also found that RAW264.7 macrophages secreted different concentrations of IL-1β and IL-10 following co-incubation with the three EcN-derived OMV types. The 50–100 nm EcN-derived OMV group showed a stronger effect in terms of inducing inflammatory cytokine secretion compared to the other two groups. This study provides direct experimental evidence that EcN-derived OMVs of different sizes exhibit heterogeneous properties.

Keywords: outer membrane vesicles; different sizes; membrane filtration; *Escherichia coli* Nissle 1917; heterogeneity; macrophage

1. Introduction

Bacterial outer membrane vesicles (OMVs) are extracellular vesicles with defined membrane structures that are actively secreted by Gram-negative bacteria [1]. Their par-

 https://doi.org/10.3390/membranes15050141

ticle size distribution is approximately 20–250 nm, and they cannot replicate [2]. OMVs contain various parent bacterial components that endow them with unique biological functions, including lipopolysaccharides, peptidoglycans, phospholipids, proteins, and nucleic acids (e.g., DNA and RNA) [3–6]. OMVs are crucial mediators of intercellular communication in bacteria–bacteria and bacteria–host interactions [7–10]. OMVs regulate various bacterial interactions, including biofilm formation, antibiotic resistance, and the killing of competing bacteria [11]. OMVs also stimulate the activities of epithelial cells, endothelial cells, and various immune cells in the host [12,13]. Proteomic analyses of OMVs provide information on their protein composition, biogenesis, and function. The gradual development of liquid chromatography with tandem mass spectrometry (LC–MS/MS) over recent years has played a pivotal role in promoting the development of proteomics research into OMVs [14–16]. Lee et al. used LC–MS/MS to demonstrate that OMVs derived from *Escherichia coli* bacteria contain various outer membrane proteins and certain porins, such as OmpA, OmpF, and NmpC. These proteins have immunostimulatory activities and can induce leukocyte migration [17]. Kwon et al. reported that OMVs derived from *Acinetobacter baumannii* contained 132 proteins, of which 43 are cytoplasmic, 26 are outer membrane proteins, 8 are inner membrane proteins, 6 are periplasmic, and the rest have an unknown localization [18]. Choi et al. reported that OMVs derived from *Pseudomonas aeruginosa* contained 338 proteins [19]. Among these proteins, the outer membrane proteins are mainly involved in the assembly of bacterial flagella, pore formation, the transport of specific substrates, and outer-membrane stability. The other proteins were found to be related to neutrophil-mediated killing.

The intestine is the largest immune organ in the body. The intestinal mucosal innate immune system, which comprises intestinal mucosal tissue, immune cells, and immune molecules, represents the body's primary line of defense against pathogenic invasion. Upon interaction with pathogenic microorganisms, the intestinal mucosal innate immune system can elicit an immune response, thereby contributing to the body's defense mechanisms [20,21]. As important innate immune cells, macrophages are among the immune cell models commonly used to study innate immune responses in vitro, and they play crucial roles in intestinal mucosal immunity. *Escherichia coli* Nissle 1917 (EcN) is a Gram-negative probiotic that resides in the human intestine. The research has indicated that EcN positively contributes to regulating host immune function and treating intestinal diseases [22]. Compared to EcN probiotics, EcN-derived OMVs exhibit a high propensity to disseminate, traverse the mucus layer, and migrate directly to the intestinal submucosal tissue to interact with host immune-system cells. Importantly, EcN-derived OMVs preserve the beneficial and immunogenic characteristics of the parent bacterium while mitigating the potential risks associated with administering live bacteria to the host. Canas et al. reported that EcN-derived OMVs mediated intestinal anti-inflammatory responses and barrier-protective effects in a mouse model of DSS-induced intestinal inflammation [23]. Fábrega et al. demonstrated that the oral administration of EcN-derived OMVs in a mouse model effectively reduced pro-inflammatory factor levels, increased the expression of intestinal barrier markers, and consequently inhibited the progression of DSS-induced colitis [24]. Alvarez et al. reported that EcN-derived OMVs activated intestinal mucosal immune and defense responses. Furthermore, these EcN-derived OMVs enhance barrier function by increasing the expression of tight junction proteins in intestinal epithelial cells [25]. There have been few studies on the heterogeneity of EcN-derived OMVs in terms of size. Isolating EcN-derived OMVs of different sizes represents the first and most important step toward conducting research regarding their heterogeneity. Currently, fast, effective, and efficient methods for the refined preparation of OMVs of different sizes are lacking. Ultrafiltration is currently used as a method for the isolation of OMVs [2]. It separates them based on their distinct physical characteristics using a combination of filter membranes of different sizes to facilitate the separation process. Inspired by the principles of ultrafiltration, we developed a membrane filtration method that is convenient and effective for separating EcN-derived OMVs of different sizes. This method has the advantages of not requiring any additional reagents, being simple to execute, being low-cost, being able to preserve the biological activity of OMVs, and being scalable for industrial mass production.

EcN-derived OMVs are pivotal in modulating intestinal immune responses. Compared to the probiotic EcN itself, EcN-derived OMVs offer unique advantages [26]. There are relatively few studies addressing the heterogeneity of EcN-derived OMVs of varying sizes, which significantly restricts research into their clinical applications. In this study, we first used the gradient filtration method to isolate high-purity OMVs derived from EcN and then conducted proteomic analyses on the isolated EcN-derived OMVs. Second, we used the filtration method to obtain EcN-derived OMVs of three specific sizes: <50 nm, 50–100 nm, and 100–300 nm. We then conducted proteomic analyses on the EcN-derived OMVs of three specific sizes. Finally, we analyzed the interactions of EcN-derived OMVs of three specific sizes with RAW264.7 macrophages to characterize their differential effects. Our results showed that the isolated EcN-derived OMVs contained a total of 646 different proteins, and the EcN-derived OMVs of different sizes contained different protein types. Among them, we found that the 50–100 nm group contained the widest variety of proteins compared to those in the <50 nm and 100–300 nm groups. We also found that RAW264.7 macrophages secreted different amounts of IL-1β and IL-10 following co-incubation with the three EcN-derived OMV types. The 50–100 nm EcN-derived OMV group showed a stronger effect in terms of inducing inflammatory cytokine secretion than the other two groups. These findings confirmed that EcN-derived OMVs of varying sizes exhibited heterogeneous characteristics.

2. Materials and Methods

2.1. *RAW264.7 Macrophage Culture*

RAW264.7 macrophages were purchased from the Shanghai Cell Biology Institute of the Chinese Academy of Sciences (Shanghai, China). Fetal bovine serum (FBS) and Dulbecco's Modified Eagle Medium (DMEM) were purchased from Thermo Fisher Scientific (Waltham, MA, USA). Penicillin-streptomycin solution was purchased from Beyotime Biotechnology (Shanghai, China). RAW264.7 macrophages were cultured in DMEM with 10% FBS and 1% penicillin-streptomycin. The cells were cultured in a CO_2 incubator with a humidified atmosphere of 5% CO_2/95% air at 37 °C.

2.2. *Isolation of EcN-Derived OMVs and Size-Based Sorting*

2.2.1. Isolation of EcN-Derived OMVs Through Gradient Filtration

The EcN bacterial stock was purchased from Baosai Biological Company (Hangzhou, China). EcN-derived OMVs were obtained via the gradient filtration protocol previously reported by our research group [27]. We optimized the protocol for the gradient filtration and significantly improved the purity of the OMV samples by repeating the pretreatment and washing steps multiple times. The detailed protocol for the isolation of EcN-derived OMVs is described below. Aliquots of EcN glycerol cryopreservation solution (100 μL) were inoculated into 10 mL of sterile LB culture medium (Sevier Biotechnology, Wuhan, China) at a 1:100 ratio and cultured at 37 °C at 200 rpm for 8 h. Subsequently, 200 μL of this seed mixture was inoculated into 200 mL of sterile LB culture medium and cultured at 37 °C at 200 rpm for 18 h. Finally, 200 mL of EcN culture medium was collected. The EcN culture medium was then centrifuged at 15,000× *g* at 4 °C to remove the bacterial cells and any sediments. The supernatant was subsequently filtered through a 0.45 μm filter (Millipore, Darmstadt, Germany) to remove residual cellular debris. The resulting supernatant was then filtered through a 0.3 μm filter membrane (Millipore). Finally, the collected filtrate was filtered and centrifuged in an ultrafiltration centrifuge tube (100 kDa, UFC9100; Millipore) at 5000× *g* until the remaining volume was approximately 0.5 mL. Residual proteins were removed via three consecutive PBS washes (12 mL of PBS solution per wash). The ultrafiltration centrifuge tube was then centrifuged at 5000× *g* for 30 min. The final precipitate (i.e., purified EcN-derived OMVs) was dissolved in 200 μL of PBS and stored at –80 °C for later use.

2.2.2. Size-Based Sorting of EcN-Derived OMVs of Varying Sizes

The EcN-derived OMV concentrate was first separated via membrane filtration with different pore sizes (100 nm, UFC30VV25; Millipore) at 5000× *g* for 5 min, with both upper solutions and lower filtrates collected to obtain <100 nm and 100–300 nm concentrates. The <100 nm concentrate was then passed through a 50 nm filter (50 nm, VMWP02500; Millipore) to separate the OMVs into <50 nm and 50–100 nm concentrates. The different OMV classes were then concentrated via ultrafiltration (100 kDa; Millipore) at 5000× *g* for 15 min.

2.3. *TEM of EcN-Derived OMVs*

TEM images were acquired using a JEM1011 electron microscope (JEOL, Tokyo, Japan). The EcN-derived OMV samples were loaded onto TEM copper grids and dried for 20 min. The samples were negatively stained with 2% phosphotungstic acid for 10 min before TEM imaging, which was performed at 80 kV.

2.4. *NTA of EcN-Derived OMVs*

NTA was performed on a NanoSight NS300 instrument (Malvern, Worchestershire, UK). The EcN-derived OMV samples were warmed to room temperature and vortexed before being serially diluted to break apart any aggregates. A 10 μL volume of the resulting mixture was then diluted with PBS to achieve a concentration ranging from 1×10^7 to 1×10^9 particles/mL. The size distribution of the OMVs was then determined via a ZetaView PMX 110 particle matrix at 405 nm.

2.5. *DLS of EcN-Derived OMVs of Varying Sizes*

DLS was performed on a Nano ZS90 instrument (Malvern). The size distribution of the EcN-derived OMVs was characterized via DLS. A 10 μL volume of the resulting mixture was diluted with 1 mL of PBS and added to a DLS test cup to perform the determination.

2.6. *BCA Protein Quantification of EcN-Derived OMVs*

A bicinchoninic acid (BCA) protein assay kit (Beyotime Biotechnology, Shanghai, China) was used to prepare a standard curve per the manufacturer's instructions. A fixed volume of each EcN-derived OMV sample was then mixed with radioimmunoprecipitation assay lysis solution. The mixture was oscillated at room temperature for 15 min. A 2 μL volume of each EcN-derived OMV sample lysate was obtained, and its absorbance (562 nm) was measured via a SpectraMax M5 full-wavelength microplate reader (Molecular Devices, Sunnyvale, CA, USA). The concentration of the EcN-derived OMVs was then determined according to the BCA standard curve and sample dilution factor.

2.7. *Purification of EcN-Derived OMV Samples via Size-Exclusion Chromatography*

Size-exclusion chromatography (SEC) was performed using the Exosupur® Purification Kit (Beijing Enzekantai Company, Beijing, China, ES914) according to the manufacturer's instructions: first, the Exosupur® exclusion column was removed from the refrigerator, maintained at 4 °C, and stabilized in an upright orientation using a test tube rack. The column was then allowed to equilibrate to ambient temperature for 30 min. Subsequently, the column was rinsed with 20 mL of PBS buffer, which was twice the column's volume. After the PBS was completely drained, 1 mL of the sample was added (if the sample was less than 1 mL, PBS was added until the total volume was 1 mL). After all samples were transferred into the sieve plate, 10 mL of the PBS eluent was added. Starting from the first drop of effluent, the effluent was collected using 1.5 mL centrifuge tubes. Each 500 µL of the effluent was marked as one fraction. Next, the OMVs were collected. It was recommended to discard fractions 1–4 and only retain fractions 5–7, as these fractions contained the highest concentration of OMVs. These collected fractions were combined into a single solution. The protein impurities predominantly appeared in the fractions collected after fraction 10, so fractions 10–15 were also collected and combined separately as a mixed protein sample. The collected OMVs and mixed protein solutions were concentrated via ultrafiltration centrifuge tube (100 kDa, UFC9100; Millipore). Subsequently, 200 µL of PBS was added to obtain an OMV concentrate and impure protein concentrate. These concentrates were subjected to quantitative analysis using the BCA protein quantification method.

2.8. SDS-PAGE of EcN-Derived OMVs

The SDS-PAGE gel preparation kit was purchased from Beyotime Biotechnology (Shanghai, China). The EcN-derived OMVs (10 µg) were boiled in 5× SDS-PAGE loading buffer for 10 min and then centrifuged (12,000× *g*), and the resulting supernatants were analyzed via SDS-PAGE on a 12% separating gel stained with G-250 protein-staining reagent (Baiolaibo Technology, Beijing, China).

2.9. Proteomic Study of EcN-Derived OMVs

Protein expression was verified using Wayen Biotechnology (Shanghai, China) via LC-MS/MS using label-free proteomic analysis, according to their established protocols [28]. Proteomic experiments with EcN-derived OMV samples were performed in biological quadruplicate. Proteomic experiments with EcN-derived OMVs of three specific sizes were performed once each. The LC-MS-MS/MS protocol is described below.

2.9.1. Preprocessing of Proteomic Samples

Equal volumes of protein lysis solution [7 M urea (Sigma, St. Louis, MO, USA), 2% SDS (Sigma), and 1% protease inhibitor cocktail (Thermo Scientific, San Jose, CA, USA)] were added to the EcN-derived OMV samples, after which an ultrasonic cell disrupter (Xinzhi Biotechnology, Ningbo, China) was used for lysis. The samples were repeatedly sonicated for 2 s, with an intervening pause of 5 s, for a total of 1 min, before being lysed on ice for 2 h. The lysed product was then centrifuged at 13,000 rpm for 20 min at 4 °C. The supernatant was collected, and its concentration was measured via BCA quantification (Beyotime Biotechnology, Shanghai, China). The samples (50 µL) of this protein mixture were then added to 10 mM dithiothreitol (DTT; Sigma) and incubated at 37 °C for 1 h. The solutions were then centrifuged at 13,000 rpm for 1 h at 4 °C to collect precipitates. The collected precipitates were added to 1 mL aliquots of 100% acetone (Thermo Scientific) for purification. The samples were then centrifuged at 13,000 rpm for 30 min, the supernatant was discarded, and the precipitate was subjected to repeated washing. The precipitate was then dried at room temperature for 10 min, dissolved in PBS, and incubated with trypsin (Promega, Sunnyvale, CA, USA) at 37 °C overnight for enzymatic digestion. Following centrifugation for 30 min at 13,000 rpm, the peptide precipitate was obtained and desalted via a C18 desalting column (GL Sciences, Tokyo, Japan).

2.9.2. LC-MS/MS Analysis

The drained peptide sample was redissolved in the mobile phase and centrifuged at 20,000× *g* for 10 min, after which the supernatant was collected for injection-processing. The peptide sample was first separated via a nanoElute UPLC liquid-phase system (Brucker, Karlsruhe, Germany). Mobile phase A consisted of 0.1% formic acid (Aladdin, Los Angeles, CA, USA) in water, and the flow rate was 300 µL/min. Mobile phase B [0.1% formic acid (Aladdin) in acetonitrile (Thermo Scientific)] was run at 2% for the first 5 min, ramped to 22% over 45 min, and ramped to 35% over 50 min, before being rapidly increased to 80% for 55 min. The peptide fragments separated by the liquid phase were ionized via the CaptiveSpray source before being input into the timsTOF Pro 2 tandem mass spectrometer (Brucker) for mode detection. The main parameters used in the mass spectrometer were as follows: ion source voltage, 1.5 kV; ion mobility range, 0.76–1.29 V·S/cm^2; first-level mass spectrometry scanning range, 452–1152 m/z; and peak intensity detected only at >2500. The 452–1152 m/z range was divided into four steps. Each step was divided into seven windows, and a total of 56 windows were used for continuous window fragmentation and information collection. The fragmentation mode was the charge injection device, and the fragmentation energy was 20–59 eV. The mass width of each window was 25. The cycle time of each DIA scan was 1.59 s.

2.9.3. Database Searching, Protein Identification, and Statistical Analyses

The obtained MS/MS data were analyzed via Spectronaut 17 (Biognosys, Schlieren, Switzerland). Database searches were performed via the UniProt database according to a previously reported protocol [29]. For proteins derived from the UniProt database, the UniProt ID was first used to match the GO ID, and the corresponding protein information was retrieved from the UniProt-GOA database on the basis of the GO ID. The following search parameters were used: enzyme digestion of Trypsin, with a parent tolerance of 15 ppm; fragment tolerance of 0.5 Da; minimum peptide length of 5; false discovery rate of 1% at the peptide and protein levels; cysteine carbamidomethylation as fixed modification; and methionine oxidation and protein N-terminal acetylation as variable modifications. The identified proteins were classified according to their subcellular localization and molecular functions via CELLO (http://cello.life.nctu.edu.tw/ (accessed on 1 September 2024)) and InterProScan (http://www.ebi.ac.uk/interpro/ (accessed on 1 September 2024)), respectively [30]. The COG protein functional classification method was used to classify the protein molecules involved in various biological processes.

2.10. Internalization of EcN-Derived OMVs of Diverse Sizes by RAW264.7 Macrophages

For the negative control group, 500 µL of PBS was combined with 0.5 µL of DiO membrane dye. In the experimental groups, 20 µg of the EcN-derived OMV samples was diluted in 500 µL of PBS and mixed with 0.5 µL of DiO membrane dye. All reaction mixtures underwent purification using an ultrafiltration centrifuge tube (100kDa) at 6000 rpm for 30 min to remove any unbound dye; two purification cycles were performed for each sample. Then, the precipitate was resuspended in 200 µL of PBS. RAW264.7 macrophages were co-cultured with the purified DiO-labeled EcN-derived OMVs of varying sizes for 16 h. The RAW264.7 macrophages were washed three times with PBS, followed by nuclear staining with Hoechst 33342 for 5 min. After washing with PBS, the RAW264.7 macrophages were added to the cell culture medium for fluorescence imaging.

2.11. Effects of EcN-Derived OMVs of Varying Sizes on Inflammatory Cytokine Secretion by RAW264.7 Macrophages

RAW264.7 macrophages were seeded into a 24-well plate at a concentration of 1×10^5/well. The cells were cultured at 37 °C and 5% CO_2 until 70–80% confluence.

A 0.95 mL aliquot of culture medium was added to each well to replace the extracted culture medium, and 50 μL aliquots of EcN-derived OMVs of different sizes at standardized concentrations of 80 μg/mL were added to final concentrations of 4 μg/mL in each well. After the RAW264.7 cells were cocultured with different sizes of EcN-derived OMVs for 24 h, the culture media were collected and centrifuged at 1500× g for 10 min at 4 °C, and the supernatants were collected. ELISA kits (Sevier Biotechnology, Wuhan, China) were used to measure the concentrations of IL-1β and IL-10 according to the manufacturer's instructions.

2.12. Statistical Analysis

Statistical analyses were performed via one-way ANOVA. All the data are presented as the mean ± SDs of three independent experiments. Significance was assessed at $p < 0.05$ (*), $p < 0.01$ (**), and $p < 0.001$ (***). All the statistical analyses were performed via GraphPad Prism software 5.0 (San Diego, CA, USA).

3. Results

3.1. EcN-Derived OMV Isolation via Gradient Filtration

We used the gradient filtration method that was previously published by our research group to separate OMVs from probiotic EcN bacteria [27]. First, the morphological characteristics of the isolated EcN-derived OMVs were analyzed via transmission electron microscopy (TEM), and their size distributions were characterized via nanoparticle tracking analysis (NTA). The morphologies of the EcN-derived OMVs were clearly observable via TEM (Figure 1a). The TEM results revealed that the isolated EcN-derived OMVs exhibited membrane structures. The NTA results revealed that the mean size of the isolated EcN-derived OMVs was 131.4 ± 3.16 nm, with a size distribution ranging from 20 to 300 nm (Figure 1b,c). These results were consistent with the previously reported structural characteristics and particle sizes of EcN-derived OMVs [31,32], confirming that the isolation procedure was successful. Next, to investigate the potential presence of protein contamination in the EcN-derived OMV samples, we used size-exclusion chromatography (SEC) to purify the samples. SEC is currently used as an effective method for achieving high-purity OMVs [2]. Our experimental results demonstrated that although we successfully acquired a concentrate of EcN-derived OMVs purified through SEC, we did not isolate a concentrated solution of the contaminating proteins. Importantly, to further confirm the purity of the EcN-derived OMVs obtained through gradient filtration, a protein analysis of the EcN-derived OMVs obtained using the two methods (SEC and gradient filtration) was carried out with SDS-PAGE. The experimental results, presented in Figure S1, demonstrated that the protein bands identified in the EcN-derived OMVs obtained using the gradient filtration corresponded to those observed via SEC. These results suggest that the purity of the EcN-derived OMVs obtained via gradient filtration is comparable to that achieved with SEC. Finally, in order to explore the therapeutic effect of EcN-derived OMVs on macrophages, we assessed the secretion of the pro-inflammatory factor IL-1β and anti-inflammatory factor IL-10 by RAW264.7 cells treated with the EcN-derived OMVs purified via SEC. As illustrated in Figure S2, we found that the RAW264.7 macrophages exhibited significantly elevated levels of IL-1β and IL-10 compared to the PBS control group (0 μg/mL of EcN-derived OMVs). Our results are consistent with those of a related study that reported that EcN-derived OMVs have the capacity to induce pro-inflammatory and anti-inflammatory responses in RAW264.7 macrophages [33]. In conclusion, the above results indicated that we efficiently isolated high-purity EcN-derived OMVs via gradient filtration.

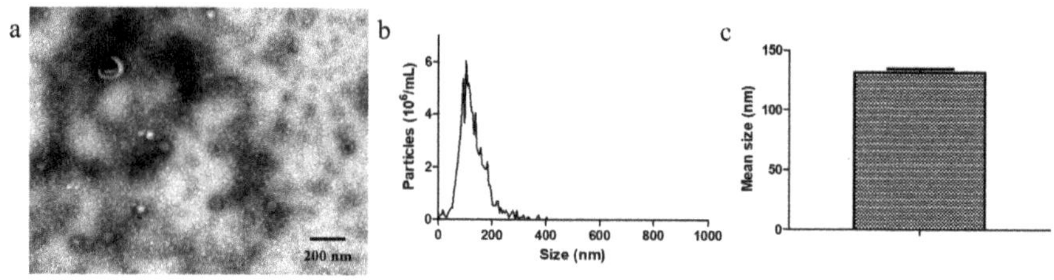

Figure 1. Transmission electron microscopy (TEM) and nanoparticle tracking analysis (NTA) of EcN-derived OMVs. (**a**) TEM images of EcN-derived OMVs (scale bar: 200 nm); (**b**) particle size distribution of EcN-derived OMVs measured via NTA; (**c**) mean size of EcN-derived OMVs. The data are shown as the mean ± SD; n = 3. The mean size refers to the average size of the EcN-derived OMVs. The mean size values were automatically generated in the NTA report.

3.2. Proteomic Analysis of EcN-Derived OMVs

We first conducted a protein analysis of the isolated EcN-derived OMVs via sodium dodecyl sulfate-polyacrylamide gel electrophoresis (SDS-PAGE) obtained from three groups of repeated experiments (Figure S3). The SDS-PAGE results revealed the abundance of proteins in the isolated EcN-derived OMVs. Importantly, in comparison to the SDS-PAGE experimental results for EcN-derived OMVs reported in the literature, the protein bands in this study were relatively singular, exhibiting a minimal presence of contaminating protein bands [26]. Furthermore, the bands of the outer membrane marker protein OmpA were prominently visible. These results further demonstrate that the EcN-derived OMVs we obtained were highly pure and exhibited no contamination from other proteins. Notably, the distribution of protein bands across the three experimental groups was consistent, indicating that the gradient filtration method used to separate EcN-derived OMVs exhibited high repeatability. Next, we performed LC-MS/MS on the isolated EcN-derived OMVs from four groups of repeated experiments, identifying 1003 proteins (Figure S4). Among these, 646 common proteins (64.4%, 646/1003) were consistently identified across all four groups. According to the subcellular locations of the 646 common identified proteins within the parent bacteria (Figure 2a and Table S1), the common proteins were classified into the following five groups: cytoplasmic (69.2%, 447/646), outer membrane (6.0%, 39/646), periplasmic (18.4%, 119/646), inner membrane (2.2%, 14/646), and extracellular (4.2%, 27/646). The subcellular locations and clusters of orthologous group (COG) functions of the 646 common proteins are summarized in Table S1. Importantly, our proteomic analysis revealed that the enriched EcN-derived OMV samples, as previously reported in the literature, contained characteristic proteins that could regulate host immune responses and provide protection against diseases. These characteristic proteins included notable outer membrane proteins (such as OmpA, OmpF, and OmpC) [33]. According to the COG functions of protein classifications (Figure 2b and Table S1), most of these common proteins were related to the biological processes of ribosomal structure/biogenesis (n = 89), the transportation and metabolism of nucleotides (n = 87), cell wall/membrane biogenesis (n = 70) and energy production and conversion (n = 69). Some proteins were also involved in transcription (n = 34); posttranslational modification and protein turnover (n = 26); replication and repair (n = 17); cycle control and cell division (n = 14); cell motility (n = 11); and the transportation and metabolism of carbohydrates (n = 28), inorganic ions (n = 24), coenzymes (n = 23), and lipids (n = 20). A few proteins related to signal transduction (n = 5), vesicular transport (n = 5), and secondary metabolite biosynthesis (n = 4) were also identified. Thus, EcN-derived OMVs contained various proteins that potentially mediated their biological functions.

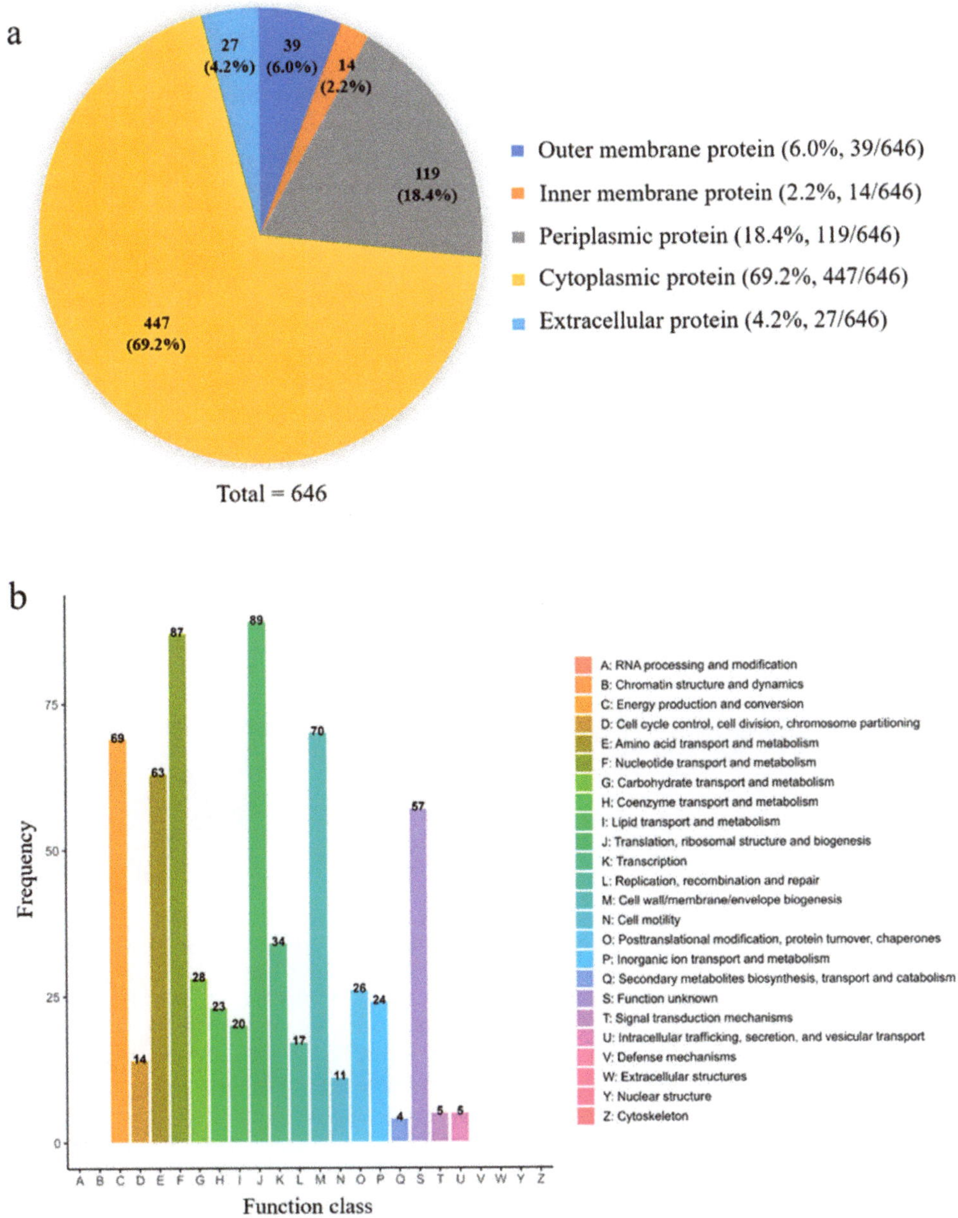

Figure 2. Proteomic analysis of EcN-derived OMVs. (**a**) Subcellular locations of the 646 common proteins identified via proteomic analysis. (**b**) We identified 646 common proteins according to a functional classification performed via the Clusters of Orthologous Groups (COG) of proteins database. The horizontal axis represents the classification content, and the vertical axis represents the relative content of the corresponding number of functional proteins.

3.3. *Segmentation of EcN-Derived OMVs of Varying Sizes and Proteomic Analysis*

The primary consideration in studying the heterogeneity of differently sized EcN-derived OMVs is how to obtain samples that meet the requirements. Currently, feasible methods for obtaining EcN-derived OMVs of different sizes are lacking. Liu et al. isolated extracellular vesicles of different sizes from cultured media with 22Rv1 prostate cancer cells via ExoTIC devices with two key pore sizes (30 and 50 nm) [34]. On the basis of this precedent, we developed a membrane filtration method to achieve the segmented separation of EcN-derived OMVs. A schematic detailing the membrane filtration method for separating OMVs of different sizes is presented in Figure S5. We used two membranes with different pore sizes (50 and 100 nm) to successfully divide the high-purity isolated EcN-derived OMVs into three distinct size groups: <50 nm, 50–100 nm, and 100–300 nm.

The procedure did not require the use of exogenous substances; therefore, the EcN-derived OMVs of three specific sizes were considered highly pure and free from contamination from other proteins. First, we conducted a TEM analysis of EcN-derived OMVs of three specific sizes. The TEM results clearly revealed the morphology of these EcN-derived OMVs (Figure 3). Notably, TEM results indicated that the three groups of EcN-derived OMVs displayed distinct size variations. Second, we used dynamic light scattering (DLS) to measure the particle size distributions of the three EcN-derived OMV classes (Figure S6). The particle size distributions of the three EcN-derived OMV classes were consistent with their respective size distribution characteristics. In conclusion, by using the membrane filtration method, we successfully obtained EcN-derived OMVs of three specific sizes: <50 nm, 50–100 nm, and 100–300 nm.

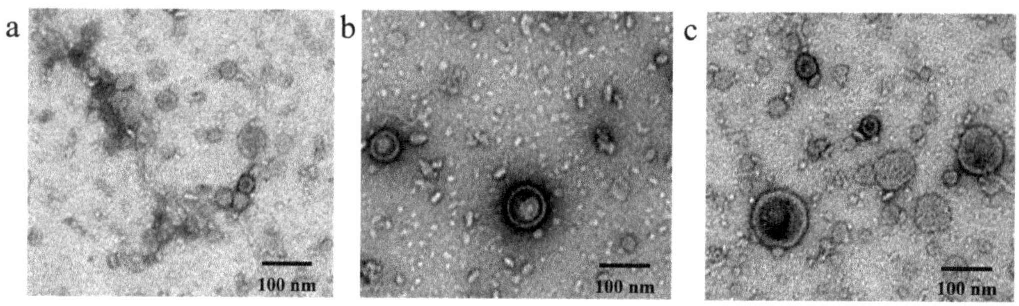

Figure 3. TEM images of EcN-derived OMVs of three specific sizes (<50 nm, 50–100 nm, and 100–300 nm): (**a**) <50 nm group; (**b**) 50–100 nm group; (**c**) 100–300 nm group. Scale bar: 100 nm.

We used LC-MS/MS analysis to perform proteomic analyses of the three groups of EcN-derived OMVs. As shown in Figure 4a and Table S2, we identified a total of 1648 unique proteins in the <50 nm, 50–100 nm, and 100–300 nm groups, comprising 15.8% (262/1648), 97.3% (1603/1648), and 95.1% (1568/1648), respectively. After analyzing these results, we found that the <50 nm group contained significantly fewer types of proteins than the 50–100 nm and 100–300 nm groups. The 50–100 nm group contained the greatest variety of proteins among the three groups. Interestingly, the percentage of common proteins among the three groups was 15.8% (261/1648). Furthermore, the 261 types of common proteins among the three groups contained five groups: cytoplasmic (115), periplasmic (84), outer membrane (36), inner membrane (4), and extracellular (22). We then focused on membrane proteins in our subsequent LC-MS/MS analyses. As shown in Figure 4b and Table S3, of the 229 total membrane proteins identified, the <50 nm, 50–100 nm, and 100–300 nm groups contributed 17.5% (40/229), 93.9% (215/229), and 91.3% (209/229), respectively. Among these membrane proteins, 17.5% (40/229) were common to both the <50 nm and 50–100 nm groups, 85.2% (195/229) were common to the 50–100 nm and 100–300 nm groups, and 17.5% (40/229) were common to the <50 nm and 100–300 nm groups. Thus, the <50 nm group contained the fewest unique membrane protein types. Finally, we analyzed the common membrane proteins contained in the three groups of

EcN-derived OMVs, as shown in Table S3. We identified 36 types of outer membrane proteins and 4 types of inner membrane proteins (including GlpT, Rne, FeoB, and BsmA) among the common membrane proteins in the three groups. Importantly, these common membrane proteins included typical proteins that could play an immunomodulatory role in host cells (such as many outer membrane proteins, including OmpA, OmpC, and OmpF). In summary, these results indicated that EcN-derived OMVs of different sizes presented obvious heterogeneity in terms of the proteins they contain.

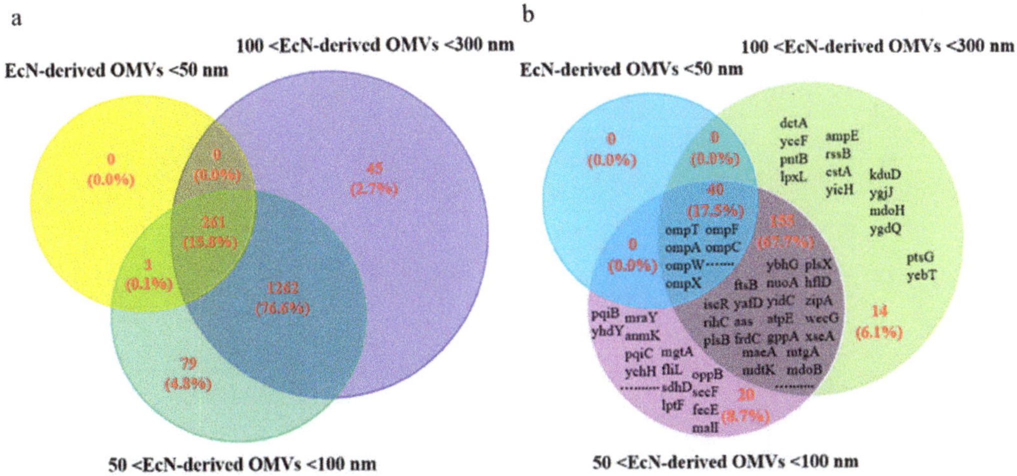

Figure 4. Proteomic analysis of the three EcN-derived OMV groups. (**a**) Venn diagram showing the unique and common proteins among the three groups of EcN-derived OMVs. (**b**) Venn diagram showing the unique and common membrane proteins among the three groups of EcN-derived OMVs.

3.4. Uptake of EcN-Derived OMVs of Varying Sizes by RAW264.7 Macrophages

In order to investigate the biological activity of EcN-derived OMVs of varying sizes, we performed a fluorescence imaging experiment in which EcN-derived OMVs of varying sizes were taken up by the RAW264.7 macrophages. As shown in Figure 5, EcN-derived OMVs of varying sizes were labeled with cell membrane dye of 3,3′-dioctadecyloxacarbocyanine perchlorate (DiO) and incubated with the RAW264.7 macrophages for 16 h. Phosphate-buffered saline (PBS) was mixed with DiO and purified through ultrafiltration to serve as the negative control group. EcN-derived OMVs of varying sizes were evaluated using fluorescence imaging to determine their ability to be taken up by the recipient RAW264.7 macrophages. As shown in Figure 5a–d, the RAW264.7 cells in the negative control group were free of DiO-labeled EcN-derived OMVs, as no fluorescence was observed in the RAW264.7 cells after 16 h of incubation. In contrast, obvious green fluorescence was observed in the RAW264.7 cells in the experimental groups after 16 h of incubation, indicating the uptake of DiO-labeled EcN-derived OMVs. These results demonstrated that EcN-derived OMVs of varying sizes can be taken up by RAW264.7 macrophages to transmit relevant molecular signals. Moreover, the uptake experiments further indicated that the EcN-derived OMVs of varying sizes showed similar biological activities. These size-independent uptake results highlighted the potential versatility of EcN-derived OMVs as natural nanocarriers for therapeutic applications.

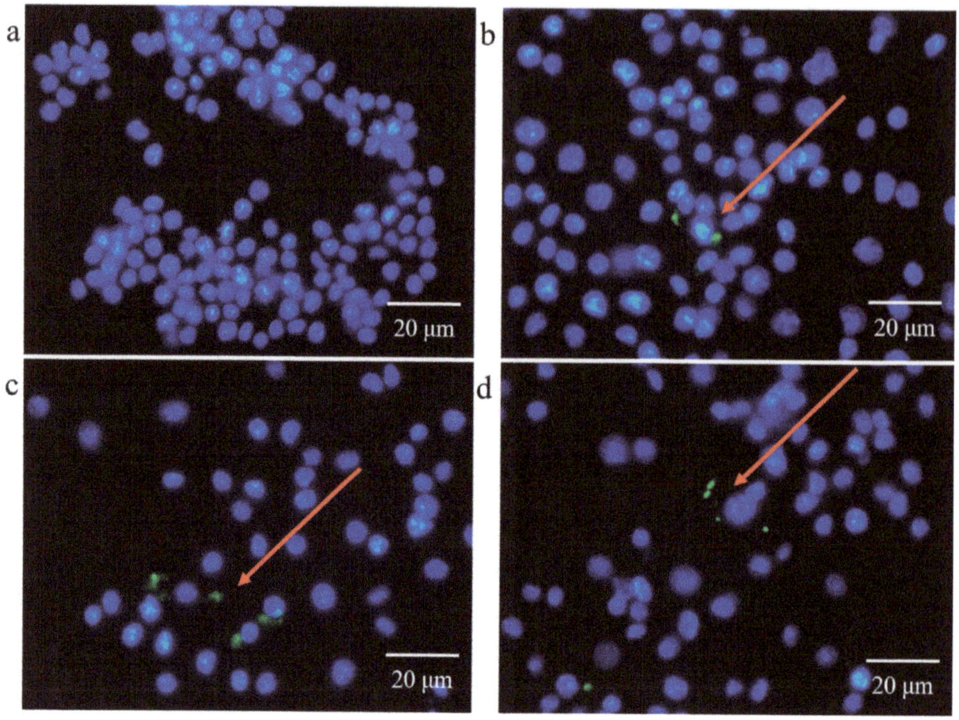

Figure 5. Fluorescence images of the RAW264.7 macrophages incubated with the DiO-labeled EcN derived OMVs of varying sizes after 16 h (scale bars: 20 μm). (**a**) Negative control (without DiO-labeled EcN-derived OMVs) and (**b**) <50 nm, (**c**) 50–100 nm, and (**d**) 100–300 nm EcN-derived OMVs labeled with DiO. The cell membrane dye DiO (green) labels EcN-derived OMVs and Hoechst 33342 (blue) labels nuclei. Representative images from three independent experiments are presented.

3.5. Effect of EcN-Derived OMVs of Varying Sizes on Inflammatory Cytokine Secretion by RAW264.7 Macrophages

Research has shown that EcN-derived OMVs can trigger both pro-inflammatory and anti-inflammatory responses in RAW264.7 macrophages [26]. To the best of our knowledge, there have been no reports on the efficacy of EcN-derived OMVs of different sizes on RAW264.7 macrophages. Thus, we co-cultured 4 μg/mL of EcN-derived OMVs of three specific sizes with RAW264.7 macrophages for 16 h and then measured the secreted pro-inflammatory IL-1β and anti-inflammatory IL-10 factors via enzyme-linked immunosorbent assay (ELISA) experiments. The standard ELISA curves are shown in Figure S7, and the results are presented in Figure 6. The experimental results indicated that the secretion level of the pro-inflammatory cytokine IL-1β by RAW264.7 macrophages in the PBS control group was minimal, whereas the anti-inflammatory cytokine IL-10 was undetectable in the cell supernatant. The levels of IL-1β produced by the experimental group of RAW264.7 macrophages were significantly greater (Figure 6a) than those produced by the PBS control group. The IL-1β concentrations in the three EcN-derived OMVs groups (<50 nm, 50–100 nm, and 100–300 nm) were 93.7 ± 8.6 pg/mL, 274.9 ± 58.1 pg/mL, and 271.9 ± 34.1 pg/mL, respectively. Furthermore, the concentration of IL-1β in the <50 nm group was lower than that in the 50–100 nm and 100–300 nm groups. Notably, the concentration of IL-10 secreted by the experimental group of RAW264.7 macrophages increased with EcN-derived OMV size (Figure 6b). The IL-10 concentration was highest in the RAW264.7 macrophages incubated with the 50–100 nm EcN-derived OMVs (67.1 ± 5.7 pg/mL), with the corresponding concentrations in those incubated with the <50 nm and 100–300 nm EcN-derived OMVs being 29.5 ± 8.6 pg/mL and 40.9 ± 2.9 pg/mL, respectively. These results suggest that EcN-derived OMVs of different sizes could induce RAW264.7 macrophages to secrete the pro-inflammatory cytokine IL-1β and the anti-inflammatory cytokine IL-10. Moreover, EcN-derived OMVs of different sizes had different effects on the ability of RAW264.7 macrophages to secrete inflammatory cytokines.

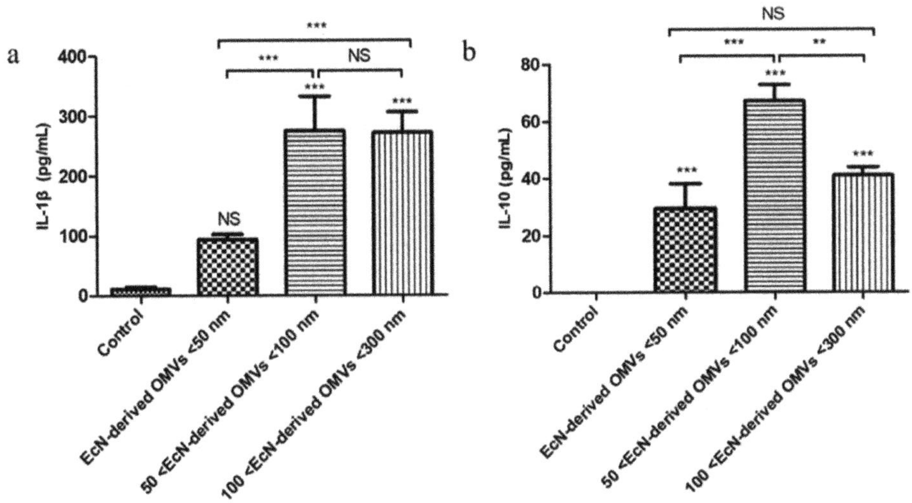

Figure 6. Effect of stimulation with the three sizes of EcN-derived OMVs at 4 μg/mL on the secretion of pro-inflammatory and anti-inflammatory cytokines by RAW264.7 macrophages: (**a**) IL-1β and (**b**) IL-10. The data are presented as the mean ± standard deviations of three independent experiments. ** $p < 0.01$, and *** $p < 0.001$, as determined via one-way analysis of variance (ANOVA).

4. Discussion

OMVs are extracellular vesicles secreted by Gram-negative bacteria with diameters ranging from 20 to 250 nm. OMVs contain biologically active molecules such as proteins, lipids, and nucleic acids, which are secreted by parent bacteria [35]. Since OMVs can transport substances derived from their parent bacteria, they may serve as effective delivery vehicles for transferring bacterial components to host cells during interactions. This mechanism has the potential to induce an immune response in the host [36]. As a novel secretion system, OMVs have many advantages. First, OMVs can selectively encapsulate the contents of parent bacteria, such as lipids, proteins, and LPS. Second, OMVs can protect their contents from being degraded by extracellular enzymes, ensuring that they can safely reach target cells. Finally, OMVs are equipped with surface ligands that interact with host cell receptors, thereby directly eliciting host immune responses. OMVs have significant potential for application in vaccine research and development, drug delivery systems, and antitumor therapeutics [2,6]. Regarding their use in vaccines, OMVs are highly immunogenic but cannot replicate; thus, they can be used as vaccine antigen candidates as well as vaccine adjuvants [37,38]. Probiotic-derived OMVs can be used as drug carriers to prevent and treat corresponding diseases; they are inanimate substances that benefit the host's health and can avoid potential harm to the host caused by live bacteria while retaining the beneficial and immunogenic properties of the parent bacteria. Regarding antitumor preparations, when combined with other treatment modalities, OMVs can enhance tumor immunotherapy and prevent tumor recurrence and metastasis [39]. EcN is a Gram-negative probiotic that is widely used in clinical treatment because of its ability to significantly alleviate intestinal diseases [40]. OMVs secreted by EcN serve as carriers of signaling molecules and can trigger pro-inflammatory and anti-inflammatory responses in RAW264.7 macrophages, thereby regulating the biological functions of these cells [26,33]. Nevertheless, the heterogeneity of EcN-derived OMVs of varying sizes has received limited scholarly attention.

Presently, frequently employed techniques for separating OMVs include ultracentrifugation, size-exclusion chromatography (SEC), ultrafiltration, and density gradient centrifugation [5,41]. Each of these methods presents distinct advantages and disadvantages. Notably, ultracentrifugation is a widely utilized technique that separates OMVs from cellular debris, bacteria, and other contaminants on the basis of the applied centrifugal force. However, this method is associated with high equipment costs, and the high centrifugal force may damage the structural integrity and functionality of the OMVs. The SEC method achieves the separation of particles on the basis of size through utilizing porous polymer gel packing. While the SEC method offers high separation purity, it is primarily suitable for the purification of samples with small volumes. In contrast, the ultrafiltration technique employs membranes with varying molecular weight cutoffs to selectively separate and collect vesicles. Ultrafiltration has the highest recovery efficiency. Density gradient centrifugation facilitates the distribution of OMVs across specific density layers via ultra-high-speed centrifugation. This method achieves the highest separation purity, but the operation is cumbersome. This study investigated the heterogeneity of EcN-derived OMVs of different sizes. The main objective was to efficiently obtain EcN-derived OMV samples of different sizes that satisfied the experimental requirements. Conventional methods fail to produce EcN-derived OMVs of different sizes. Inspired by the experimental principles of ultrafiltration, we developed a membrane filtration method to obtain EcN-derived OMVs of varying sizes. In this study, we first used the gradient filtration previously developed by our research group specifically for the separation of EcN-derived OMVs, and used SEC to assess the purity of the EcN-derived OMV samples. The experimental findings indicated that the purity of the EcN-derived OMVs obtained via gradient filtration was comparable to that achieved through SEC. Our approach to optimizing the gradient filtration method, through the incorporation of pretreatment steps and the repeated washing of the samples, proved to be highly effective. We then used the membrane filtration method to separate the high-purity isolated EcN-derived OMVs into three major size groups (<50 nm, 50–100 nm, and 100–300 nm), which were confirmed using TEM and DLS. The significant advantages of the membrane filtration method are that it is fast, convenient, and suitable for achieving industrial-scale yields for mass production, and it maintains the biological functional activity of the isolated OMVs.

OMVs contain biologically active molecules, including proteins, lipids, polysaccharides, DNA, and RNA derived from parent bacterial cells. Elucidating the molecular composition of OMVs is crucial for comprehending their function in bacterial growth or interaction with the host. The analysis of OMV components mainly employs proteomic technologies to elucidate their protein composition. Proteomic research on the OMVs of Gram-negative pathogenic bacteria has identified many protein constituents associated with bacterial pathogenesis and immune modulation [42,43]. In this study, our proteomics results revealed that the isolated EcN-derived OMVs contained 646 total protein types. There is a paucity of published research concerning proteomic analyses of EcN-derived OMVs of different sizes. Nice et al. reported that the *Aggregatibacter actinomycetemcomitans* strain JP2 produces two types of vesicles throughout growth, with varying sizes and toxin compositions [44]. Turner et al. reported that the size of *Helicobacter pylori*-derived OMVs determines their protein content, as fewer and less diverse bacterial proteins are contained within smaller OMVs than those contained within larger OMVs [45]. In addition, there are related reports on the heterogeneity of EVs derived from animal cells. Extracellular vesicles (EVs) released by tumor cells have received considerable attention in recent years, as they contain biomarkers such as tumor-specific proteins and nucleic acids (i.e., RNA and DNA) [46,47]. Liu et al. researched the effect of EV size on protein profiles. They isolated EVs of different sizes from cultured media with 22Rv1 prostate cancer cells via ExoTIC devices with two key pore sizes (30 and 50 nm), and reported that EVs of different sizes contained different protein types [34]. In this study, we used LC-MS/MS analysis to perform proteomic analyses of EcN-derived OMVs of three specific sizes: <50 nm, 50–100 nm, and 100–300 nm. Our results revealed that the <50 nm group contained significantly fewer protein types (262) than the 50–100 nm (1603) and 100–300 nm (1568) groups did. We conducted a categorical analysis of 261 proteins commonly present in EcN-derived OMVs of three specific sizes. They were classified into five groups: cytoplasmic (44.1%, 115/261), periplasmic (32.2%, 84/261), outer membrane (13.8%, 36/261), inner membrane (1.5%, 4/261), and extracellular (8.4%, 22/261). Given that the EcN-derived OMVs of three specific sizes were all secreted by the parent bacteria, it was entirely plausible that they shared certain common proteins. The results indicated that the EcN-derived OMVs of three specific sizes possessed a relatively rich variety of common proteins, which suggested that they might perform similar biological functions. As we found in this study, the EcN-derived OMVs of three specific sizes had obvious therapeutic effects on RAW264.7 macrophages and significantly promoted the secretion of the pro-inflammatory factor IL-1β and the anti-inflammatory factor IL-10 by RAW264.7 macrophages. Although our study provides limited proteomic data on EcN-derived OMVs of three distinct sizes, these experimental findings substantiate that EcN-derived OMVs of varying sizes present differences in protein composition. Future research will involve a comprehensive analysis of proteomic data from EcN-derived OMVs of different sizes, with a focus on investigating the types and characteristics of these differential proteins in greater depth.

Most inflammatory responses result from innate immunity, which is mainly caused through the activation of pro-inflammatory signaling cascades by macrophages and leukocytes. The inflammatory cytokines produced by this cascade comprise small soluble proteins produced by these immune cells; these cytokines regulate the host's immune and inflammatory responses. Inflammatory cytokines involved in inflammatory responses are mainly divided into two categories, namely, pro-inflammatory factors (e.g., TNF-α, IL-6, and IL-1β) and anti-inflammatory factors (e.g., IL-4 and IL-10) [48]. Among them, IL-1β can stimulate immune cells to secrete pro-inflammatory factors such as TNF-α, which plays a pro-inflammatory role. TNF-α is an important regulatory factor in the initiation of immunity and inflammation. IL-10 is an anti-inflammatory factor that can play an anti-inflammatory role. EcN-derived OMVs can stimulate RAW264.7 macrophages to secrete the pro-inflammatory factors TNF-α, IL-6, and IL-1β, and the anti-inflammatory factor IL-10 [26]. In this study, the concentrations of the pro-inflammatory cytokine IL-1β produced by RAW264.7 macrophages in the three types of EcN-derived OMVs of different sizes (<50 nm, 50–100 nm, and 100–300 nm) were 93.7 $\pm$ 8.6 pg/mL, 274.9 $\pm$ 58.1 pg/mL, and 271.9 $\pm$ 34.1 pg/mL, respectively. Moreover, the concentration of anti-inflammatory IL-10 produced by RAW264.7 macrophages in the 50–100 nm group was the highest (67.1 $\pm$ 5.7 pg/mL), and the concentrations of IL-10 in the <50 nm and 100–300 nm groups were 29.5 $\pm$ 8.6 pg/mL and 40.9 $\pm$ 2.9 pg/mL, respectively. Our results revealed that the therapeutic effects of EcN-derived OMVs of different sizes differed; the 50–100 nm group had a stronger therapeutic effect than the other two groups. We analyzed the reasons for the different experimental phenomena observed regarding the macrophage function mentioned above and speculate that these differences may be attributable to the distinct protein compositions of EcN-derived OMVs of varying sizes. Proteomic analyses of the

EcN-derived OMVs of three specific sizes revealed that, in addition to the 261 common proteins shared among the three groups, the 50–100 nm group contained 1341 unique proteins that were distinct from those in the <50 nm group. Similarly, the 100–300 nm group contained 1307 unique proteins that were distinct from those in the <50 nm group. These experimental results suggest that the 50–100 nm and 100–300 nm groups might have stronger therapeutic effects on RAW264.7 macrophages than the <50 nm group. Furthermore, a comparative analysis between the 50–100 nm group and the 100–300 nm group revealed that, in addition to sharing 1523 common proteins, each group also contained distinct proteins. The 50–100 nm group contained 80 unique proteins, whereas the 100–300 nm group contained 45 unique proteins. This variation in protein composition within each group might account for the differential therapeutic effects observed between the 50–100 nm group and the 100–300 nm group on RAW264.7 macrophages. In summary, these differences in macrophage function are consistent with the differences in the proteomic information of the EcN-derived OMVs of varying sizes that we measured. In the macrophage function experiment, we only measured changes in the representative pro-inflammatory cytokine IL-1β and the anti-inflammatory cytokine IL-10. Subsequent experiments should be conducted to further analyze the changes in other pro-inflammatory factors, such as TNF-α, IL-6, and other anti-inflammatory factors, such as IL-4.

Although our experimental findings suggest that EcN-derived OMVs of varying sizes are heterogeneous in terms of protein composition and interactions with RAW264.7 macrophages, additional research is necessary to establish more comprehensive conclusions. Our preliminary proteomic results suggest that EcN-derived OMVs of different sizes present distinct protein compositions. The composition of other components in EcN-derived OMVs of varying sizes, such as lipopolysaccharides, phospholipids, DNA, and RNA, is currently unknown. This study aimed to elucidate the differences in the therapeutic effects of EcN-derived OMVs of varying sizes on RAW264.7 macrophages through examining the distinct protein compositions associated with these EcN-derived OMVs. Importantly, this may represent only one contributing factor. The variations in other constituents contained in EcN-derived OMVs of different sizes may also influence their effects on RAW264.7 macrophages. Further investigation into these unidentified components is necessary in future research. In addition, the therapeutic efficacy of EcN-derived OMVs of varying sizes on macrophages in mouse models remains unclear and requires further validation. However, notwithstanding these limitations, our results provide direct experimental evidence that EcN-derived OMVs of different sizes exhibit heterogeneous properties.

5. Conclusions

EcN-derived OMVs play a crucial role in modulating intestinal immune responses and possess significant potential for clinical applications. Few studies have addressed the heterogeneity of differently sized EcN-derived OMVs, significantly limiting research on their clinical applications. In this study, we successfully obtained EcN-derived OMVs of three specific sizes (<50 nm, 50–100 nm, and 100–300 nm) via the membrane filtration method. Among them, the <50 nm group (262) contained fewer types of total proteins than the 50–100 nm (1603) and 100–300 nm (1568) groups. Moreover, in terms of the differential secretion of the pro-inflammatory factor IL-1β and the anti-inflammatory factor IL-10 in RAW264.7 macrophages, the 50–100 nm group presented a stronger effect than the other two groups. Collectively, our findings reveal that the size of EcN-derived OMVs plays a key role in regulating macrophage function and their protein cargo composition. These findings underscore a significant concern regarding the clinical application of research on EcN-derived OMVs, as variations in their size may influence the experimental outcomes. We speculate that variations may exist not only in their protein composition but also in other constituent elements, as well as in the pathways and mechanisms through which EcN-derived OMVs of different sizes interact with the host. Therefore, we conclude that EcN-derived OMVs of varying sizes exhibit heterogeneous characteristics. These findings have fundamental and significant implications that should be considered when examining the role of EcN-derived OMVs in clinical applications.

Supplementary Materials: The following supporting information can be downloaded at: https://www.mdpi.com/article/10.3390/membranes15050141/s1. Supplementary File S1 (docx): Figure S1: Protein analysis of the *Escherichia coli* Nissle 1917 (EcN)-derived OMVs obtained using two different methods [size-exclusion chromatography (SEC) and gradient filtration] was carried out with sodium dodecyl sulfate–polyacrylamide gel electrophoresis (SDS-PAGE). Figure S2: Effect of stimulation with the EcN-derived OMVs purified via SEC (4 μg/mL) on the secretion of pro-inflammatory and anti-inflammatory cytokines by RAW264.7 macrophages. Figure S3: Protein analysis of EcN-derived OMVs was conducted via SDS-PAGE, with three groups of repeated experiments. Figure S4: Venn diagram showing the 646 common proteins between the four groups of repeated experiments. Figure S5: Schematic of the segmentation of OMVs into three types of different sizes (<50 nm, 50–100 nm, and 100–300 nm). Figure S6: The particle size distributions of the differently sized EcN-derived OMVs were measured via dynamic light scattering (DLS): (a) <50 nm EcN-derived OMVs; (b) 50–100 nm EcN-derived OMVs; (c) 100–300 nm EcN-derived OMVs. Figure S7: Standard curves were used to determine the pro-inflammatory and anti-inflammatory cytokine concentrations via ELISA: standard curves for (a) IL-1β and (b) IL-10. Table S1: EcN-derived OMV proteins identified in this study. Supplementary File S2 (xlsx): Table S2: Comparison of the total proteins contained in the EcN-derived OMVs of three specific sizes (<50 nm, 50–100 nm, and 100–300 nm). Supplementary File S3 (xlsx): Table S3: Comparison of the membrane proteins contained in the EcN-derived OMVs of three specific sizes (<50 nm, 50–100 nm, and 100–300 nm).

Author Contributions: Conceptualization, N.L. and K.D.; investigation, N.L. and K.D.; data curation, N.L. and K.D.; writing—original draft preparation, N.L. and K.D.; writing—review and editing, N.L., H.X. and K.D.; resources and funding acquisition, N.L., H.X. and K.D. All authors have read and agreed to the published version of the manuscript.

Funding: This work was supported by the National Natural Science Foundation of China (grant number: 82360055 to N.L.), the National Key Research and Development Program of China (grant number: 2022YFA1104300 to K.D. and H.X.), and the Jiangxi Provincial Natural Science Foundation (grant numbers: 20232BAB216005 and 20242BAB25427 to N.L.).

Institutional Review Board Statement: Not applicable.

Informed Consent Statement: Not applicable.

Data Availability Statement: The MS proteomics data have been deposited in the ProteomeXchange Consortium (http://proteomecentral.proteomexchange.org (accessed on 1 December 2024)) via the iProX partner repository with the dataset identifier PXD058454.

Conflicts of Interest: The authors declare no competing interests.

Abbreviations

OMVs	Outer membrane vesicles
LC-MS/MS	Liquid chromatography with tandem mass spectrometry
EcN	*Escherichia coli* Nissle 1917
TEM	Transmission electron microscopy
NTA	Nanoparticle tracking analysis
DLS	Dynamic light scattering
SDS-PAGE	Sodium dodecyl sulfate-polyacrylamide gel electrophoresis
BCA	Bicinchoninic acid
DiO	3,3′-dioctadecyloxacarbocyanine perchlorate

IL-1β	Interleukin-1β
IL-10	Interleukin-10
ELISA	Enzyme-linked immunosorbent assay
PBS	Phosphate-buffered saline
EVs	Extracellular vesicles
SEC	Size-exclusion chromatography

References

1. Kim, J.H.; Lee, J.; Park, J.; Gho, Y.S. Gram-negative and Gram-positive bacterial extracellular vesicles. *Semin. Cell Dev. Biol.* **2015**, *40*, 97–104. [CrossRef]
2. Xie, J.; Li, Q.; Haesebrouck, F.; Van Hoecke, L.; Vandenbroucke, R.E. The tremendous biomedical potential of bacterial extracellular vesicles. *Trends Biotechnol.* **2022**, *40*, 1173–1194. [CrossRef]
3. Vanaja, S.K.; Russo, A.J.; Behl, B.; Banerjee, I.; Yankova, M.; Deshmukh, S.D.; Rathinam, V.A.K. Bacterial Outer Membrane Vesicles Mediate Cytosolic Localization of LPS and Caspase-11 Activation. *Cell* **2016**, *165*, 1106–1119. [CrossRef]
4. Bhar, S.; Edelmann, M.J.; Jones, M.K. Characterization and proteomic analysis of outer membrane vesicles from a commensal microbe, Enterobacter cloacae. *J. Proteom.* **2021**, *231*, 103994. [CrossRef] [PubMed]
5. Li, M.; Zhou, H.; Yang, C.; Wu, Y.; Zhou, X.; Liu, H.; Wang, Y. Bacterial outer membrane vesicles as a platform for biomedical applications: An update. *J. Control. Release* **2020**, *323*, 253–268. [CrossRef] [PubMed]
6. Chen, H.; Zhou, M.; Zeng, Y.; Lv, Z.; Wang, P.; Han, L. Recent advances in biomedical applications of bacterial outer membrane vesicles. *J. Mater. Chem. B* **2022**, *10*, 7384–7396. [CrossRef]
7. Rueter, C.; Bielaszewska, M. Secretion and Delivery of Intestinal Pathogenic Escherichia coli Virulence Factors via Outer Membrane Vesicles. *Front. Cell. Infect. Microbiol.* **2020**, *10*, 91. [CrossRef] [PubMed]
8. Cañas, M.-A.; Fábrega, M.-J.; Giménez, R.; Badia, J.; Baldomà, L. Outer Membrane Vesicles From Probiotic and Commensal Escherichia coli Activate NOD1-Mediated Immune Responses in Intestinal Epithelial Cells. *Front. Microbiol.* **2018**, *9*, 498. [CrossRef]
9. Uddin, M.J.; Dawan, J.; Jeon, G.; Yu, T.; He, X.; Ahn, J. The Role of Bacterial Membrane Vesicles in the Dissemination of Antibiotic Resistance and as Promising Carriers for Therapeutic Agent Delivery. *Microorganisms* **2020**, *8*, 670. [CrossRef]
10. Oh, J.; Ñahui Palomino, R.A.; Vanpouille, C.; Costantini, P.E.; Margolis, L. Microbiota–host communications: Bacterial extracellular vesicles as a common language. *PLOS Pathog.* **2021**, *17*, e1009508.
11. Lee, J.; Kim, O.Y.; Gho, Y.S. Proteomic profiling of Gram-negative bacterial outer membrane vesicles: Current perspectives. *PROTEOMICS—Clin. Appl.* **2016**, *10*, 897–909. [CrossRef]
12. Ellis, T.N.; Kuehn, M.J. Virulence and Immunomodulatory Roles of Bacterial Outer Membrane Vesicles. *Microbiol. Mol. Biol. Rev.* **2010**, *74*, 81–94. [CrossRef]
13. Fiocca, R.; Necchi, V.; Sommi, P.; Ricci, V.; Telford, J.; Cover, T.L.; Solcia, E. Release of Helicobacter pylori vacuolating cytotoxin by both a specific secretion pathway and budding of outer membrane vesicles. *Uptake Released Toxin Vesicles By Gastric Ep. J. Pathol.* **1999**, *188*, 220–226.
14. Zhang, K.; Chu, P.; Song, S.; Yang, D.; Bian, Z.; Li, Y.; Gou, H.; Jiang, Z.; Cai, R.; Li, C. Proteome Analysis of Outer Membrane Vesicles From a Highly Virulent Strain of Haemophilus parasuis. *Front. Vet. Sci.* **2021**, *8*, 756764. [CrossRef] [PubMed]
15. Zhang, X.; Wang, Y.; Fan, R.; Zhang, L.; Li, Z.; Zhang, Y.; Zheng, W.; Wang, L.; Liu, B.; Quan, C. Quantitative Proteomic Analysis of Outer Membrane Vesicles from Fusobacterium nucleatum Cultivated in the Mimic Cancer Environment. *Microbiol. Spectr.* **2023**, *11*, e0039423. [CrossRef]
16. Wang, Y.; Deng, J.; Wang, X.; Zhang, L.; Xu, Y.; Ren, J.; Niu, S.; Zhao, Y.; Yan, F.; Tian, W.-x.; et al. Isolation, identification, and proteomic analysis of outer membrane vesicles of Riemerella anatipestifer SX-1. *Poult. Sci.* **2024**, *103*, 103639. [CrossRef] [PubMed]
17. Lee, E.Y.; Bang, J.Y.; Park, G.W.; Choi, D.S.; Kang, J.S.; Kim, H.J.; Park, K.S.; Lee, J.O.; Kim, Y.K.; Kwon, K.H.; et al. Global proteomic profiling of native outer membrane vesicles derived from Escherichia coli. *Proteomics* **2007**, *7*, 3143–3153. [CrossRef]
18. Kwon, S.-O.; Gho, Y.S.; Lee, J.C.; Kim, S.I. Proteome analysis of outer membrane vesicles from a clinicalAcinetobacter baumannii-isolate. *FEMS Microbiol. Lett.* **2009**, *297*, 150–156. [CrossRef]
19. Choi, D.S.; Kim, D.K.; Choi, S.J.; Lee, J.; Choi, J.P.; Rho, S.; Park, S.H.; Kim, Y.K.; Hwang, D.; Gho, Y.S. Proteomic analysis of outer membrane vesicles derived from Pseudomonas aeruginosa. *Proteomics* **2011**, *11*, 3424–3429. [CrossRef]
20. Macpherson, A.J.; Harris, N.L. Interactions between commensal intestinal bacteria and the immune system. *Nat. Rev. Immunol.* **2004**, *4*, 478–485. [CrossRef]
21. Magalhaes, J.G.; Tattoli, I.; Girardin, S.E. The intestinal epithelial barrier: How to distinguish between the microbial flora and pathogens. *Semin. Immunol.* **2007**, *19*, 106–115. [CrossRef] [PubMed]

22. Trebichavsky, I.; Splichal, I.; Rada, V.; Splichalova, A. Modulation of natural immunity in the gut by Escherichia coli strain Nissle 1917. *Nutr. Rev.* **2010**, *68*, 459–464. [CrossRef] [PubMed]
23. Omri, A.; Cañas, M.-A.; Giménez, R.; Fábrega, M.-J.; Toloza, L.; Baldomà, L.; Badia, J. Outer Membrane Vesicles from the Probiotic Escherichia coli Nissle 1917 and the Commensal ECOR12 Enter Intestinal Epithelial Cells via Clathrin-Dependent Endocytosis and Elicit Differential Effects on DNA Damage. *PLoS ONE* **2016**, *11*, e0160374.
24. Fábrega, M.-J.; Rodríguez-Nogales, A.; Garrido-Mesa, J.; Algieri, F.; Badía, J.; Giménez, R.; Gálvez, J.; Baldomà, L. Intestinal Anti-inflammatory Effects of Outer Membrane Vesicles from Escherichia coli Nissle 1917 in DSS-Experimental Colitis in Mice. *Front. Microbiol.* **2017**, *8*, 1274. [CrossRef] [PubMed]
25. Alvarez, C.-S.; Giménez, R.; Cañas, M.-A.; Vera, R.; Díaz-Garrido, N.; Badia, J.; Baldomà, L. Extracellular vesicles and soluble factors secreted by Escherichia coli Nissle 1917 and ECOR63 protect against enteropathogenic E. coli-induced intestinal epithelial barrier dysfunction. *BMC Microbiol.* **2019**, *19*, 166. [CrossRef]
26. José Fábrega, M.; Aguilera, L.; Giménez, R.; Varela, E.; Alexandra Cañas, M.; Antolín, M.; Badía, J.; Baldomà, L. Activation of Immune and Defense Responses in the Intestinal Mucosa by Outer Membrane Vesicles of Commensal and Probiotic Escherichia coli Strains. *Front. Microbiol.* **2016**, *7*, 705. [CrossRef]
27. Li, N.; Wu, M.; Wang, L.; Tang, M.; Xin, H.; Deng, K. Efficient Isolation of Outer Membrane Vesicles (OMVs) Secreted by Gram-Negative Bacteria via a Novel Gradient Filtration Method. *Membranes* **2024**, *14*, 135. [CrossRef] [PubMed]
28. Ma, S.; Xing, X.; Huang, H.; Gao, X.; Xu, X.; Yang, J.; Liao, C.; Zhang, X.; Liu, J.; Tian, W.; et al. Skeletal muscle-derived extracellular vesicles transport glycolytic enzymes to mediate muscle-to-bone crosstalk. *Cell Metab.* **2023**, *35*, 2028–2043.e7. [CrossRef]
29. Aguilera, L.; Toloza, L.; Giménez, R.; Odena, A.; Oliveira, E.; Aguilar, J.; Badia, J.; Baldomà, L. Proteomic analysis of outer membrane vesicles from the probiotic strain Escherichia coli Nissle 1917. *Proteomics* **2014**, *14*, 222–229. [CrossRef]
30. Hu, R.; Li, J.; Zhao, Y.; Lin, H.; Liang, L.; Wang, M.; Liu, H.; Min, Y.; Gao, Y.; Yang, M. Exploiting bacterial outer membrane vesicles as a cross-protective vaccine candidate against avian pathogenic Escherichia coli (APEC). *Microb. Cell Factories* **2020**, *19*, 119. [CrossRef]
31. Shi, J.; Ma, D.; Gao, S.; Long, F.; Wang, X.; Pu, X.; Cannon, R.D.; Han, T.-L. Probiotic Escherichia coli Nissle 1917-derived outer membrane vesicles modulate the intestinal microbiome and host gut-liver metabolome in obese and diabetic mice. *Front. Microbiol.* **2023**, *14*, 1219763. [CrossRef]
32. Han, L.; Liu, X.-W.; Li, W.-X.; Zang, T.; Li, Z.-P.; Deng, Y.; Tan, L.-L.; Liang, D.-S.; Zhang, X.-J.; Shen, X.-F.; et al. Innovative probiotic *Escherichia coli* nissle 1917-derived outer membrane vesicles coupled microspheres demonstrate anti-inflammatory and barrier-enhancing properties in colitis. *Chem. Eng. J.* **2024**, *487*, 150711. [CrossRef]
33. Hu, R.; Lin, H.; Li, J.; Zhao, Y.; Wang, M.; Sun, X.; Min, Y.; Gao, Y.; Yang, M. Probiotic *Escherichia coli* Nissle 1917-derived outer membrane vesicles enhance immunomodulation and antimicrobial activity in RAW264.7 macrophages. *BMC Microbiol.* **2020**, *20*, 268. [CrossRef] [PubMed]
34. Liu, F.; Vermesh, O.; Mani, V.; Ge, T.J.; Madsen, S.J.; Sabour, A.; Hsu, E.-C.; Gowrishankar, G.; Kanada, M.; Jokerst, J.V.; et al. The Exosome Total Isolation Chip. *ACS Nano* **2017**, *11*, 10712–10723. [CrossRef]
35. Brown, L.; Wolf, J.M.; Prados-Rosales, R.; Casadevall, A. Through the wall: Extracellular vesicles in Gram-positive bacteria, mycobacteria and fungi. *Nat. Rev. Microbiol.* **2015**, *13*, 620–630. [CrossRef]
36. Kaparakis-Liaskos, M.; Ferrero, R.L. Immune modulation by bacterial outer membrane vesicles. *Nat. Rev. Immunol.* **2015**, *15*, 375–387. [CrossRef] [PubMed]
37. Wo, J.; Lv, Z.-Y.; Sun, J.-N.; Tang, H.; Qi, N.; Ye, B.-C. Engineering probiotic-derived outer membrane vesicles as functional vaccine carriers to enhance immunity against SARS-CoV-2. *Iscience* **2023**, *26*, 105772. [CrossRef]
38. Chen, L.; Valentine, J.L.; Huang, C.-J.; Endicott, C.E.; Moeller, T.D.; Rasmussen, J.A.; Fletcher, J.R.; Boll, J.M.; Rosenthal, J.A.; Dobruchowska, J. Outer membrane vesicles displaying engineered glycotopes elicit protective antibodies. *Proc. Natl. Acad. Sci. USA* **2016**, *113*, E3609–E3618. [CrossRef]
39. Gujrati, V.; Kim, S.; Kim, S.-H.; Min, J.J.; Choy, H.E.; Kim, S.C.; Jon, S. Bioengineered bacterial outer membrane vesicles as cell-specific drug-delivery vehicles for cancer therapy. *ACS Nano* **2014**, *8*, 1525–1537. [CrossRef]
40. Scaldaferri, F.; Gerardi, V.; Mangiola, F.; Lopetuso, L.R.; Pizzoferrato, M.; Petito, V.; Papa, A.; Stojanovic, J.; Poscia, A.; Cammarota, G.; et al. Role and mechanisms of action of Escherichia coliNissle 1917 in the maintenance of remission in ulcerative colitis patients: An update. *World J. Gastroenterol.* **2016**, *22*, 5505. [CrossRef]
41. Tulkens, J.; De Wever, O.; Hendrix, A. Analyzing bacterial extracellular vesicles in human body fluids by orthogonal biophysical separation and biochemical characterization. *Nat. Protoc.* **2019**, *15*, 40–67. [CrossRef] [PubMed]
42. Haurat, M.F.; Elhenawy, W.; Feldman, M.F. Prokaryotic membrane vesicles: New insights on biogenesis and biological roles. *Biol. Chem.* **2015**, *396*, 95–109. [CrossRef] [PubMed]
43. Watanabe, K. Bacterial membrane vesicles (MVs): Novel tools as nature- and nano-carriers for immunogenic antigen, enzyme support, and drug delivery. *Appl. Microbiol. Biotechnol.* **2016**, *100*, 9837–9843. [CrossRef] [PubMed]

44. Nice, J.B.; Collins, S.M.; Agro, S.M.J.; Sinani, A.; Moros, S.D.; Pasch, L.M.; Brown, A.C. Heterogeneity of Size and Toxin Distribution in Aggregatibacter actinomycetemcomitans Outer Membrane Vesicles. *Toxins* **2024**, *16*, 138. [CrossRef]
45. Turner, L.; Bitto, N.J.; Steer, D.L.; Lo, C.; D'Costa, K.; Ramm, G.; Shambrook, M.; Hill, A.F.; Ferrero, R.L.; Kaparakis-Liaskos, M. Helicobacter pylori Outer Membrane Vesicle Size Determines Their Mechanisms of Host Cell Entry and Protein Content. *Front. Immunol.* **2018**, *9*, 1466. [CrossRef]
46. Syn, N.L.; Wang, L.; Chow, E.K.-H.; Lim, C.T.; Goh, B.-C. Exosomes in Cancer Nanomedicine and Immunotherapy: Prospects and Challenges. *Trends Biotechnol.* **2017**, *35*, 665–676. [CrossRef]
47. Sheridan, C. Exosome cancer diagnostic reaches market. *Nat. Biotechnol.* **2016**, *34*, 359. [CrossRef]
48. Feghali, C.A.; Wright, T.M. Cytokines in acute and chronic inflammation. *Front. Biosci. A J. Virtual Libr.* **1997**, *2*, d12.

membranes

MDPI

Article

Efficient Separation of a Novel Microbial Chassis, *Vibrio natriegens*, from High-Salt Culture Broth Using Ceramic Ultrafiltration Membranes

Stefan Schwarz [1], Rong Fan [1,2,3], Mehrdad Ebrahimi [1] and Peter Czermak [1,4,*]

1 Institute of Bioprocess Engineering and Pharmaceutical Technology, University of Applied Sciences Central Hesse, 35390 Giessen, Germany; stefan.schwarz@lse.thm.de (S.S.); fanrong@mail.ipe.ac.cn (R.F.); mehrdad.ebrahimi@lse.thm.de (M.E.)
2 State Key Laboratory of Biopharmaceutical Preparation and Delivery, Institute of Process Engineering, Chinese Academy of Sciences, Beijing 100190, China
3 School of Chemical Engineering, University of Chinese Academy of Sciences, Beijing 100190, China
4 Faculty of Biology and Chemistry, Justus-Liebig-University Giessen, 35390 Giessen, Germany
* Correspondence: peter.czermak@lse.thm.de

Abstract: *Vibrio natriegens* is widely used as a production host for biotechnological processes due to its superior maximum glucose consumption rate, high growth rate, and abundant ribosomes. Most bioprocesses also need a scalable biomass separation step. This can be achieved by cross-flow filtration with ceramic membranes, although the membrane pores are susceptible to fouling. However, the fouling characteristics of *V. natriegens* culture broth have not been investigated in detail. We therefore characterized membrane fouling during the separation of *V. natriegens* biomass from culture broth using a cross-flow filtration plant with ceramic membranes. The resistance in series model was used to quantify the fouling-induced resistance caused by the different components of the culture broth. The total fouling resistance was $4.1 \cdot 10^9 \pm 0.6 \cdot 10^9$ m^{-1} for the culture broth and $5.4 \cdot 10^9 \pm 0.7 \cdot 10^9$ m^{-1} for the summed broth components. Reversible resistance accounted for 86% and 81% of these totals, respectively. We then applied Hermia's adapted filtration laws to determine the dominant fouling mechanism induced by the different broth components. In a further step, we established a setup to determine the compressibility index of the cells during cross-flow filtration, resulting in an estimated value of 0.55 ± 0.04. These results will facilitate the design of economic filtration plants and will help to establish *V. natriegens* as a production host for large-scale industrial processes.

Keywords: fouling; *Vibrio natriegens*; compressibility index; resistance in series model; Hermia's laws; ceramic membranes; ultrafiltration; CFF

1. Introduction

Vibrio natriegens is a rod-shaped, halophilic, Gram-negative marine bacterium with an approximate size of 3.5×0.3 µm [1,2]. Its superior maximum glucose consumption rate and high growth rate make it useful for many biotechnological applications, which are facilitated by its compatibility with the plasmid systems originally developed for *Escherichia coli* [3–6]. Accordingly, *V. natriegens* has been developed as an expression host for the production of recombinant peptides and proteins and has also been used to develop a cell-free expression system and functional nanoparticles [7–15]. Owing to its high salt tolerance, *V. natriegens* can be cultivated under a non-sterile condition, thus decreasing the operating costs for sterilization in large-scale industrial processes. As a promising chassis

https://doi.org/10.3390/membranes15040121

in synthetic biology, *V. natriegens* is able to secrete organic alcohols, acids, and heterologous proteins as products into the extracellular space. Like other cell-based expression systems, it is necessary to separate the *V. natriegens* biomass from the culture broth at the beginning of downstream processing. This can be achieved by cross-flow filtration with ceramic membranes, but the fouling characteristics of the culture broth have not been investigated in detail, and this is the starting point for process optimization.

Ceramic membranes are widely used for the filtration of organic process streams in the food and feed, pulp and paper, and bulk chemical industries, as well as in the biotechnology sector for the purification of proteins [16–22]. The advantages of ceramic membranes include their resistance to harsh conditions, such as extreme pH and temperatures, as well as chemical cleaning agents. This means they can be regenerated frequently, which is beneficial because membrane fouling is a major limitation to the efficiency of filtration processes. Other advantages include their high resistance to abrasive media, their longevity, and their high flux per unit area, which combine to make them more attractive from an economic perspective [23–25]. Owing to their high chemical stability, ceramic membranes are suitable to separate *V. natriegens* with the products from a high-salt-containing medium. However, this hypertonic environment may change the surface interaction between biological particles and membrane materials and cause severe fouling, which changes the separation efficiency and selectivity. Therefore, we systematically analyzed the fouling mechanism during the filtration of *V. natriegens* in a hypertonic solution with ceramic membranes.

The flux through a membrane depends on the driving force, the total resistance, and the permeate viscosity. The driving force during filtrations with porous membranes is the transmembrane pressure. The total resistance is composed of the membrane resistance and the fouling resistance. The membrane resistance can be calculated using Equation (1), which is based on Darcy's law [26,27]. The membrane resistance is equal to the total resistance when the permeate is pure water passing through a clean membrane.

$$R_m = R_T = \frac{\Delta p}{\mu * J} \tag{1}$$

where R_m is the membrane resistance, R_T is the total resistance, Δp is the transmembrane pressure, μ is the dynamic viscosity, and J is the filtrate flux.

During filtration, a filter cake builds up on the retentate side of the membrane, and the total resistance is therefore the sum of the membrane resistance and cake resistance. In this case, the flux can be calculated using Equation (2) [28].

$$J = \frac{\Delta p}{\mu \cdot (\alpha \cdot w + R_m)} \tag{2}$$

where α is the specific cake resistance and w is the weight of the cake per filtration area. The specific cake resistance is calculated using Equation (3), which is a rearrangement of Equation (2).

$$\alpha = \frac{\Delta p /_{J \cdot \mu} - R_m}{w} \tag{3}$$

We can differentiate between the true specific cake resistance and the apparent specific cake resistance; the latter is shown in Equation (3). The true specific cake resistance is calculated by replacing R_m in Equation (3) with R_{mi}, the sum of pristine membrane resistance and resistance caused by the fouling inside pores, resulting in Equation (4) [26]. Several approaches have been described to determine the cake mass. In addition to weighing the entire membrane or the mechanically removed cake, it is also possible to remove the membrane and digest the cake to deduce the cake mass from the amount of protein released [29–31]. Another approach, which makes it possible to determine the cake mass without removing the membrane, is to measure the biomass in the feed and calculate the cake mass as the difference before and after filtration [26].

$$\alpha = \frac{\Delta p / J \cdot \mu - R_{mi}}{w} \tag{4}$$

To understand R_{mi}, it is necessary to know that the resistances that lead to membrane fouling can be divided into those reflecting reversible and irreversible fouling. Reversible fouling is the portion of fouling that can be removed by mechanical cleaning, which can involve procedures such as backwashing, gas scouring, and relaxation, whereas irreversible fouling cannot be removed. The irreversible fouling resistance of a membrane is determined by passing water through a mechanically cleaned membrane. The value is then calculated using Equation (1) but replacing R_m with R_{mi}. The reversible fouling resistance R_{rf} is the difference between the total resistance of the fouled membrane minus the resistance of the cleaned membrane and the irreversible fouling resistance R_{if}, resulting in R_{mi} and leading to Equation (5) [25,27,32,33]. The distinction between reversible and irreversible fouling allows us to assign specific fouling mechanisms to each of the resistances. Irreversible resistance is caused by the fouling mechanisms of complete blocking, standard blocking, and, in part, intermediate blocking, whereas reversible resistance is caused by cake deposition combined with intermediate blocking (Figure 1).

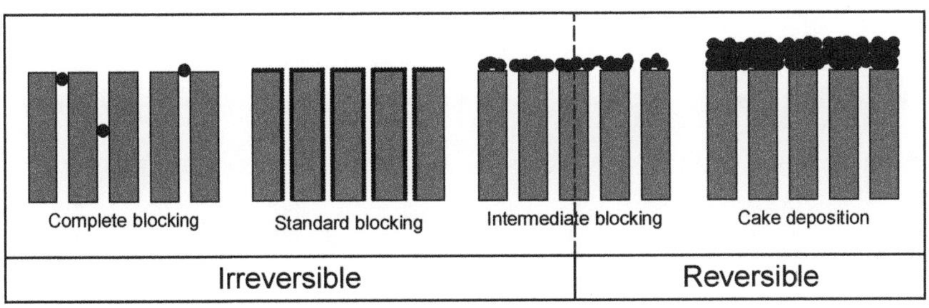

Figure 1. Visualization of different fouling mechanisms, adapted from Di Bella and Di Trapani (2019) [27].

In addition to the aforementioned definitions, another definition of reversible fouling includes all fouling that can be removed mechanically and chemically, whereas irreversible fouling includes fouling that cannot be removed by these means [34,35]. These alternative definitions are not used in the current article.

$$R_T = R_m + R_{if} + R_{rf} \tag{5}$$

With increasing transmembrane pressure, the specific cake resistance also increases. This can be mathematically expressed using Equation (6), where α' is a coefficient and n is the compressibility index [28,36,37]. Several factors affect the compressibility index, including cell shape and size, as well as the ionic strength of the feed [28,38,39]. Ellipsoidal cells tend to have lower compressibility indices than rod-shaped cells [28,39].

$$\alpha = \alpha' \cdot (\Delta p)^n \tag{6}$$

The rearrangement of Equation (6) gives Equation (7). These equations can be solved for n by measuring α at two different transmembrane pressures (Equation (8)).

$$\alpha' = \frac{\alpha}{(\Delta p)^n} \quad (7)$$

$$n = log_{\frac{\Delta p_2}{\Delta p_1}}\left(\frac{\alpha_2}{\alpha_1}\right) \quad (8)$$

The term Δp is often used for the transmembrane pressure [28,38,40], but the force driving compression is not transmembrane pressure but the proportion of the pressure that falls on the cake [41]. This proportion depends on the individual resistances of the cake and the membrane, and it changes with the membrane resistance if the cake resistance remains constant. For this reason, we prefer to use the pressure drop across the cake to calculate the compressibility index so that it can be considered and compared independently of the membrane [42]. The pressure drop over the cake is calculated using Equation (9).

$$\Delta p_c = \Delta p - \mu \cdot J \cdot R_{mi} \quad (9)$$

There are two popular approaches to describe fouling mechanisms. One is the resistance in series model, which distinguishes components in the feed and allocates to each a share of the total resistance. Equation (10) formalizes this approach [43,44]. According to this model, the sum of the fouling resistances of the medium (R_{ME}), cells (R_C), and byproducts of microbial activity (R_P) is equal to the fouling resistance of the fermentation broth.

$$R_{FB} = (R_{ME} + R_C + R_P) \quad (10)$$

The other approach distinguishes between the fouling mechanisms of complete blocking, standard blocking, intermediate blocking, and cake deposition. The mathematical description of this model for dead-end filtration is based on Equation (11) [27,45,46]. The mechanisms underlying the different types of fouling (and their reversibility, if appropriate) are summarized in Figure 1.

$$\frac{d^2t}{dV^2} = k\left(\frac{dt}{dV}\right)^n \quad (11)$$

The different fouling mechanisms are also known as Hermia's laws [47], and the corresponding Equations (12)–(15) are presented in Table 1.

Table 1. Hermia's laws for dead-end filtration.

Mechanism	n	Equation	
Complete blocking	2	$J = J_0 \cdot exp(-k \cdot J_0 \cdot t)$	(12)
Standard blocking	1.5	$J = \frac{J_0}{(1+J_0 \cdot k \cdot t)^2}$	(13)
Intermediate blocking	1	$J = \frac{J_0}{1+J_0 \cdot k \cdot t}$	(14)
Cake deposition	0	$J = \frac{J_0}{(1+J_0 \cdot k \cdot t)^{0.5}}$	(15)

This model was adapted to cross-flow filtration by including the influence of convective transport on the fouling layer (Equation (16)). Equations (17)–(20) in Table 2 distinguish the different fouling mechanisms [47,48]. In comparison to the other mechanisms, standard blocking is assumed to take place generally inside the membrane pores; this fact is dominated by the diffusion of solutes passing through the membrane pores. It suggests that standard blocking is not influenced by the cross-flow rate.

$$-\frac{dJ}{dt}(J^{n-2}) = k(J - J_{ss}) \quad (16)$$

Table 2. Hermia's laws for cross-flow filtration.

Mechanism	n	Equation	
Complete blocking	2	$J = J_{ss} + (J_0 - J_{ss})\cdot exp(-k\cdot t)$	(17)
Standard blocking	1.5	$\frac{1}{J^{0.5}} = \frac{1}{J_0^{0.5}} + k\cdot t$	(18)
Intermediate blocking	1	$k\cdot t = \frac{1}{J_{ss}}\cdot \ln\left(\frac{J}{J_0}\cdot\frac{J_0 - J_{ss}}{J\cdot J_{ss}}\right)$	(19)
Cake deposition	0	$k\cdot t = \frac{1}{J_{ss}^2}\left(\ln\left(\frac{J_0}{J}\cdot\frac{J_0-J_{ss}}{J-J_{ss}}\right)\right) - J_{ss}\cdot\left(\frac{1}{J} - \frac{1}{J_0}\right)$	(20)

2. Materials and Methods

2.1. Production of Fermentation Broth

We cultivated *V. natriegens* Vmax cells (Synthetic Genomics, Calipatria, CA, USA) in 250 mL of the recommended medium [7] in 1 L baffled shake flasks (Schott, Mainz, Germany). The medium was inoculated at an OD_{600} of 0.03 and cultivated overnight in a Multitron shaking incubator (Infors, Bottmingen, Switzerland) set at 32 °C and 250 rpm.

2.2. Feed Solutions for Filtration with the Ceramic Membrane

The feed solutions tested in the resistance in series model were *V. natriegens* cells in 3% NaCl, medium, culture supernatant, and culture broth. The medium deviated from the medium used for fermentation by excluding glucose. All feed solutions were adjusted to pH 7.4 ± 0.1 using 1 M NaOH or 1 M HCl. To adjust the OD_{600} of the culture broth, a portion was centrifuged at 8000 g for 10 min, and the culture broth was diluted with the supernatant to an OD_{600} of 7.4. To prepare the cell solutions in 3% NaCl, 1.4 L of the overnight culture was centrifuged as described above, and the pellets were resuspended in 600 mL of 3% NaCl and then centrifuged again. The pellets were then resuspended in 3% NaCl, and the suspension was diluted in 3% NaCl until the OD_{600} was 7.4 ± 5% (for the fouling models) or 0.95 ± 5% (for the compressibility index experiments). The supernatant was generated by single-step centrifugation of the overnight culture, as described above, followed by passage through a 0.2 µm bottle-top filter (VWR, Radnor, PA, USA).

2.3. Filtration Plant

The filtrations tested using the resistance in series model and the fits using Hermia's laws were based on a UF50A membrane (Atech Innovations, Gladbeck, Germany) with a 50 nm cut-off, an inner diameter of 6 mm, and an effective length of 23.5 cm, resulting in a filtration area of 0.0044 m^2. The filtrations for the compressibility index tests were carried out using the same type of membrane, an effective length of 45 cm, an outer diameter of 25.4 cm, and 19 channels, each with a diameter of 3.3 mm, resulting in a filtration area of 0.089 m^2. The membrane materials and support materials consisted of Al_2O_3. The construction of the filtration plant is shown in Figure 2. The membrane was inserted into a stainless-steel holder, and the feed from the tempered double-jacket reservoir was directed through an SM6000 flow sensor (IFM, Essen, Germany) to the membrane via an FCPA 80B-4/HE rotary vane pump (AFT, Rosstal, Germany). The feed temperature was adjusted by cooling the reservoir using an ECO RE 420 thermostat (Lauda, Lauda-Koenigshofen, Germany). Pressure sensors were located before and behind the membrane. The pressure sensor behind the membrane was adjacent to a ball valve, allowing the transmembrane pressure to be adjusted. Behind the ball valve, the feed was recirculated back to the feed reservoir. On the permeate side of the membrane, the permeate was fed into a beaker via a pressure sensor and a downstream ball valve. The beaker was placed on a PFB 3000-2 electrical scale (Kern & Sohn, Balingen-Frommern, Germany) linked to a laptop. The permeate flow was thus recorded by averaging over 10 s using LabVision (HiTec Zang, Herzogenrath, Germany). The permeate in the beaker was emptied back into the permeate reservoir before the mass in the beaker exceeded 10% of the feed mass.

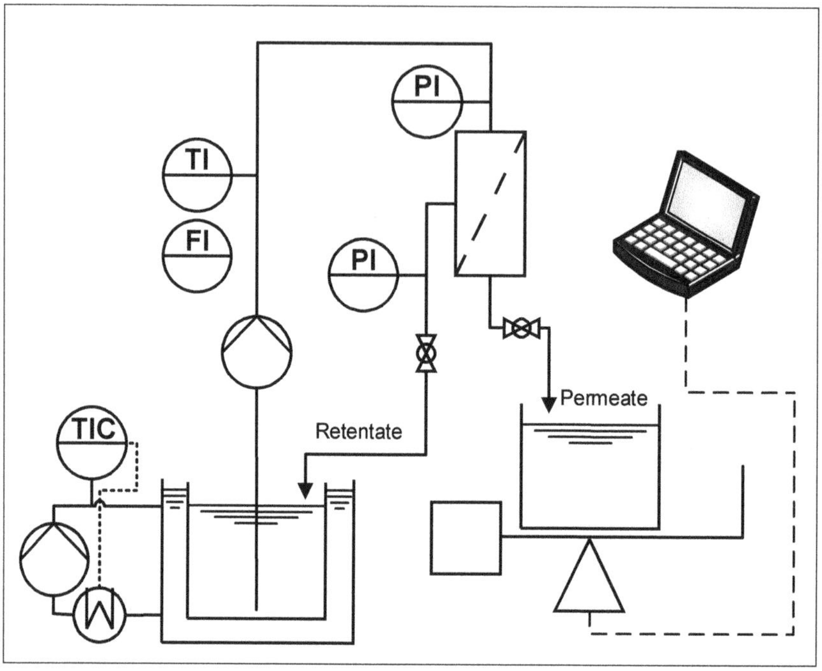

Figure 2. Setup of the filtration plant: PI = pressure indicator, TI = temperature indicator, FI = flow indicator, and TIC = temperature indicator and controller.

2.4. General Procedure for the Filtration Experiments

Before each filtration experiment, the cleaning solution in the filtration system was replaced with water until the permeate pH became neutral. The membrane resistance was then determined by measuring the permeate flux at five equidistant transmembrane pressures from 0.4 to 2.0 bar, using Equation (1). In the next step, the water was drained and replaced with 3% NaCl. Afterward, a feed volume of 1200 ± 100 mL was filled into the system and circulated for 10 min under filtration conditions with a closed permeate valve. Filtrations for the resistance in series model were carried out for a minimum of 90 min or until a steady-state transmembrane flux was established, whereas filtrations to determine the compressibility index were completed after 20 min. For each filtration testing the resistance in series model, a transmembrane pressure of 0.8 bar was applied, while the temperature was maintained at 25 ± 1 °C. For the single-channel membrane, we used a cross-flow velocity of 3 $m \cdot s^{-1}$, whereas a higher volumetric flow, equivalent to a flow velocity value of 0.5 $m \cdot s^{-1}$, was used for the 19-channel membrane. This corresponds to a wall shear rate of approximately 1212 s^{-1} for the 19-channel membrane (3.3 mm channel diameter) and 4000 s^{-1} for the single-channel membrane (6 mm channel diameter), calculated based on the respective flow velocities. Samples were taken from the permeate and feed to determine the protein concentration using a Bradford assay, as described by Schwarz et al. [7]. After filtration, we evaluated the viscosity of a permeate sample using an MCR102 modular compact rheometer (Anton Paar, Ostfildern-Scharnhausen, Germany). The feed was drained and replaced with at least 1 L of water, which was circulated within the filtration plant at a cross-flow velocity of 4.5 $m \cdot s^{-1}$ (single-channel membrane) or

$0.8\ m{\cdot}s^{-1}$ (19-channel membrane) for 15 min with a closed permeate valve. The water was then replaced, and the procedure was repeated with an open permeate valve. Having replaced the feed with fresh water again, the resistance of the irreversibly fouled membrane was determined as described above.

2.5. Filtration Experiments to Determine the Compressibility Index

The general procedure described above was followed with the following deviations. After rinsing the cleaning solution, the water was replaced with 3% NaCl. The cross-flow velocity was set to $0.5\ m{\cdot}s^{-1}$, and the transmembrane pressure was set to the value used for subsequent filtration. The feed was circulated in the filtration system with the permeate valve closed using the specified parameters for 10 min before filtration was initiated by opening the permeate valve. To avoid any influence on the biomass concentration in the feed, the permeate flow was only determined at the end of the filtration; otherwise, the permeate was returned directly to the feed reservoir. Five filtrations were performed at transmembrane pressures of 0.4–2.0 bar in equidistant steps. A feed sample was taken every minute to determine the OD_{600}. At the end of filtration, the permeate flux was measured three times, and the viscosity of a permeate sample was determined as described above.

2.6. Chemical Cleaning of Ceramic Membranes

The membranes were cleaned at the same cross-flow velocities applied during filtration and at a transmembrane pressure of 0.4 bar. The first cleaning solution was $8\ g{\cdot}L^{-1}$ Dismozon (Hartmann, Heidenheim, Germany), applied for 1 h. The second cleaning solution was 1% (*w*/*v*) citric acid (Carl Roth, Karlsruhe, Germany), applied for 1 h. The last cleaning solution was 1% (*w*/*v*) P3 Ultrasil 14 (Ecolab Deutschland, Monheim am Rhein, Germany), applied for 2 h and also used for the subsequent storage of the filtration plant.

2.7. Data Analysis

The resistances of the clean and irreversibly fouled membranes were calculated using Equation (1). The water fluxes of the clean membrane were used to calculate the membrane resistance. The total fouling resistance was calculated using the flux of the fouled membrane minus the value generated with the clean membrane. The reversible fouling resistance was the difference between the total fouling resistance and the irreversible resistance. The irreversible fouling resistance was the resistance of the mechanically cleaned membrane minus the resistance of the clean membrane. To compare resistances for the resistance in series model, total fouling resistance values were calculated from the flux 90–95 min after the start of filtration. All fitting was carried out using Origin (Origin Lab, Northampton, MA, USA). The initial value for the compressibility index fit was calculated using Equation (8). The corresponding specific cake resistances were calculated using Equations (6) and (9). The cake mass was calculated using the OD_{600} values. Fouling analysis according to Hermia's laws was carried out using Equations (12)–(20).

3. Results

3.1. Resistance in Series Model

The resistance in series model was used to determine the contribution of different components in the culture broth to the fouling resistance. The resistances of the medium, the cell suspension, and the culture broth were determined directly via the permeate flows before and after filtration, as well as after mechanical cleaning, as described in the Materials and Methods Section. The resistances of the microbial byproducts result indirectly from the subtraction of the resistances of the supernatant and the medium. Figure 3 provides an overview of the procedure for collecting the necessary data.

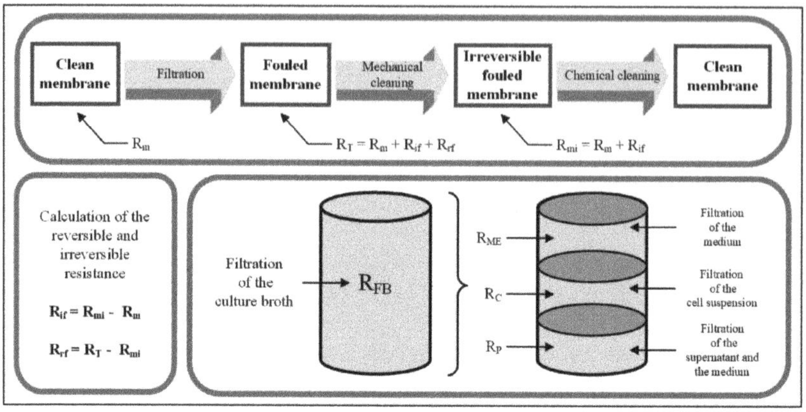

Figure 3. Overview of the procedure for calculating the resistances of the resistance in series model.

Microbial culture broth is a complex mixture consisting of cells, cell debris, released biomacromolecules, and small molecular substances. Owing to the difference in size and surface chemistry, these bioparticles may exhibit various transmembrane manners. Our previous study demonstrated that large particles like cells can be completely rejected by a 50 nm ceramic membrane, while proteins were partially adsorbed in the inner pores of the membranes. The content of these components alters the flux and predominant fouling mechanism during the filtration. In order to figure out the individual influence of each factor on fouling formation, we analyzed the fouling resistances by filtrating different bio-fluids step by step.

Microbial byproducts accounted for the largest share of the total resistance (48%), followed by the cells (47%) and the medium (5%). The sum of these individual resistances was compared to the resistance of the culture broth, and the same comparison was carried out for the reversible and irreversible resistances. For the total resistance and reversible resistance, the sum of the individual resistances overlapped with that of the culture broth when taking experimental uncertainty into account, but for the irreversible resistance, the sum of individual resistances was higher than that of the culture broth (Figure 4). This can be explained by cell lysis in NaCl during filtration. We compared the OD_{600} of cell suspensions in 0.9% and 3% NaCl solutions. It remained stable in the higher concentration of NaCl but gradually declined in the 0.9% NaCl solution, suggesting cell lysis and loss in viability caused by low osmolarity. The osmolarity in the medium is 1.41 mol/L, and the osmolarity of 30 g/L NaCl is lower at 1.03 mol/L. A solution with 9 g/L NaCl has an even lower osmolarity of 0.34 mol/L. All microorganisms, except perhaps halophilic *Halobacteria*, have a positive turgor. In order to maintain this osmotic pressure, they must adapt the osmolarity of the cell interior to their environment. To this end, bacteria vary the concentration of salts and compatible solutes inside the cell [49]. *Vibrionaceae*, which include *Vibrio natriegens*, are able to adapt their osmolarity through the uptake and synthesis of compatible solutes [50]. If the osmolarity in the environment decreases faster than in the cell, the turgor increases, which can lead to a weakening of the cell and disruption [51]. Another possible explanation is the absence of substrates or the combination of both circumstances, considering that the transport of compatible solutes across the cell wall is partly dependent on ATP. The protein concentration in the cell suspension increased between the 30 min and 90 min time points during filtration, while the reverse was observed during the filtration of culture broth (Figure 5). Here, a notable fact is that the protein concentration in the permeate of the culture broth was higher than it was in the feed of cell-free supernatant, suggesting a release of protein from cells during the filtration of the culture broth (Figure 5B,D). We assume that the harsh conditions within the filtration system led to the desorption of the cell-bound proteins that remained associated with the cells during the preparation of the supernatant. This hypothesis is further supported by the significant protein concentration observed in the permeate during the filtration of cells suspended in a NaCl solution.

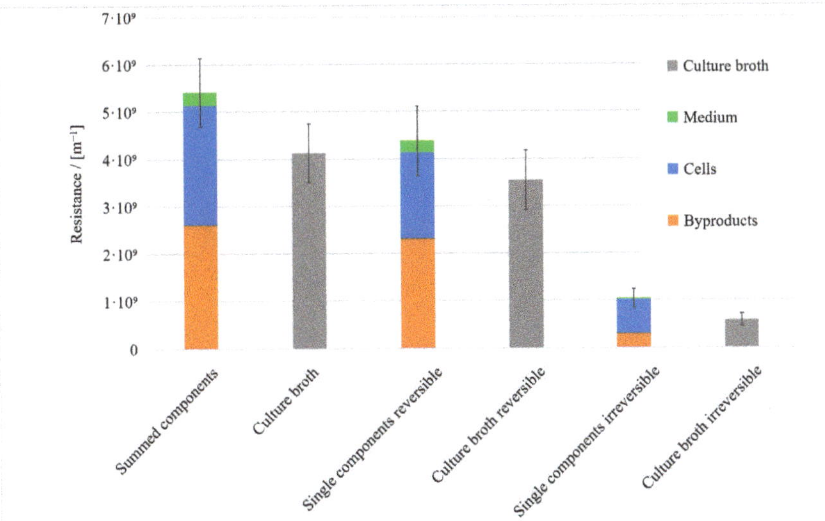

Figure 4. Resistances of broth components compared to the overall resistance of the culture broth. Error bars were calculated by Gaussian error propagation.

Considering the biological effects described above, the lysis of *V. natriegens* observed in the preliminary experiments (see the supplemental data) led to our hypothesis that cell debris was the cause of the increase in irreversible fouling. The decreasing protein concentration in the culture supernatant may reflect the adsorption of proteins onto the membrane and/or depletion of the proteins due to the metabolic activity of *V. natriegens*. Members of the genus *Vibrio* are known to secrete metalloproteases and serine proteases [52]. In addition, numerous amino acids are among the substrates that can be utilized by *V. natriegens* [53]. This may also explain why this continuous degradation did not occur in the cell suspension. Due to the cell separation and subsequent wash step, the proteases were separated from the cell suspension. The constant protein concentration during the filtration of the supernatant excludes the first hypothesis, making the second the more likely explanation. The protein concentration in the permeate of the supernatant filtration is lower than in the feed. We assume that this is due to the fact that some of the proteins are retained by the membrane. This conclusion is also consistent with the observation that in the case of the microbial byproducts, a part of the fouling resistance is of a reversible nature. The initial increase in the protein concentration in the culture broth in the period of 5 to 30 min after the start of filtration may be a result of cell lysis caused by the shear forces in the pump. This increase cannot be observed in the cell suspension. This can be attributed to the shear forces during the centrifugation of the cells when preparing the cell suspension. As a result, a more shear-sensitive proportion of the bacterial population may have already been lysed before the cell suspension was used in the filtration system.

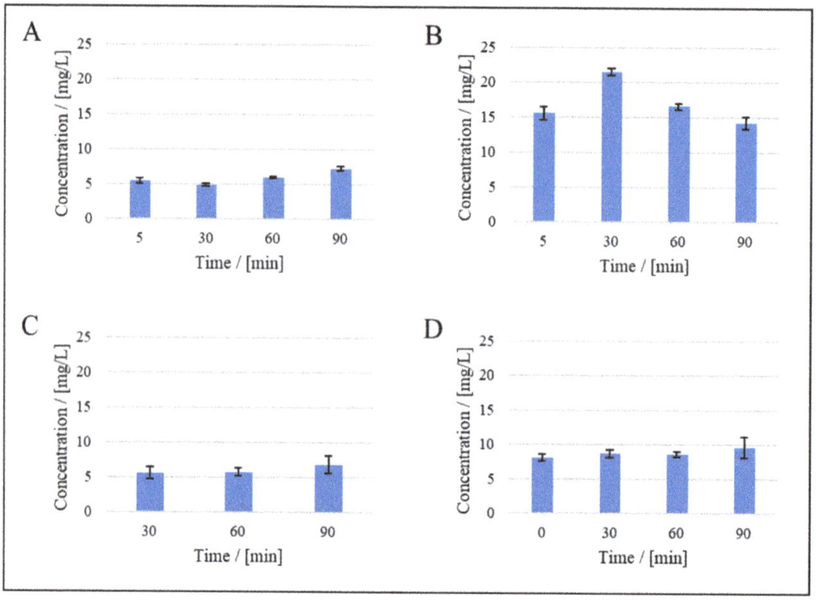

Figure 5. Protein concentration in the feed and permeate: (**A**) = cells, permeate side, (**B**) = culture broth, permeate side, (**C**) = supernatant, permeate side, and (**D**) = supernatant, feed side. Data are means ± standard deviations (n = 3).

3.2. Compressibility Index

The compressibility index was calculated using shorter filtration periods to ensure that cell lysis had a negligible impact on the fouling behavior. In the preliminary experiments, the OD_{600} of the feed declined while circulating in the filtration system. We assume this was caused by cell lysis during filtration under suboptimal conditions, which increased the degree of irreversible fouling by the cell lysate. This is supported by the increasing protein concentration between the 30 min and 90 min time points (Figure 5). The preliminary experiments revealed that compressibility was affected by the culture age and NaCl concentration. Therefore, cells were taken from fresh overnight cultures, and we set the NaCl concentration to 30 $g \cdot L^{-1}$ in the feed. The filtration time was also reduced to 20 min, resulting in a smaller proportion of irreversible fouling relative to the total fouling resistance (1–3%). In the longer filtration runs for the other fouling models, the proportion of the irreversible fouling resistance was 29%. The proportion of irreversible fouling in total fouling resistance appears to be variable. A closer examination with the aid of the errors reveals that the irreversible resistance at 1.6 bar cannot be explained by the errors (see the supplemental data). We assume that this is an outlier.

The measurement of cell mass in the feed during these experiments ensured that an equilibrium was reached between the mass of the cells in the feed and in the cake on the membrane ~15 min after the start of filtration (Figure 6). Similar findings, in which an equilibrium was established within 10–15 min, were reported in another study [26]. The authors also observed continual flux decline after cake formation, which agrees with our observation of irreversible fouling (Figure 7). This was attributed, in part, to the medium components remaining after the cell wash [26], in agreement with the protein concentrations we measured 5 min after the start of filtration (Figure 5).

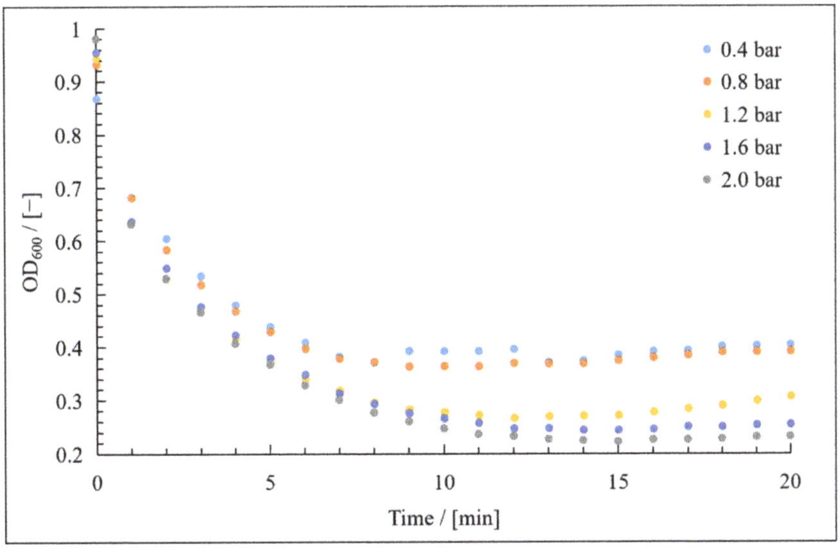

Figure 6. OD_{600} values measured during the course of filtration to determine the compressibility index.

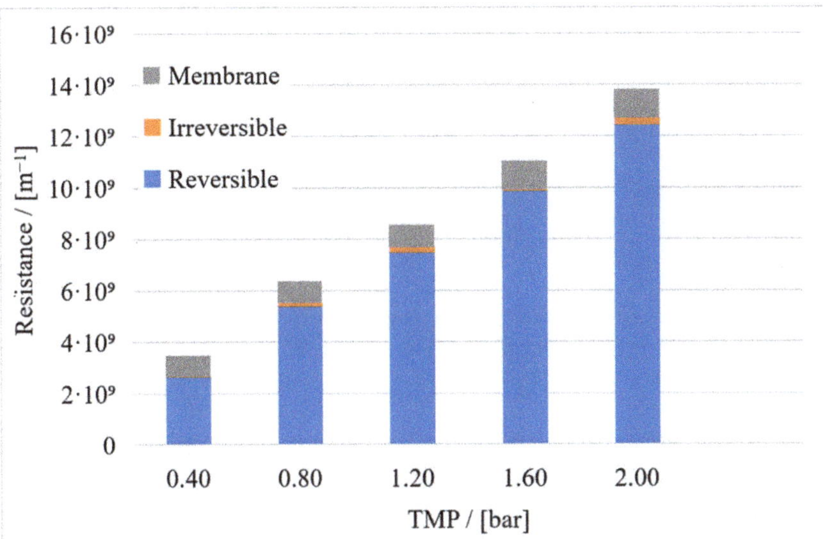

Figure 7. Components of total resistance to calculate the compressibility index (resistance of the cleaned membrane, and resistances attributed to reversible and irreversible fouling).

The first calculation of the compressibility index, based on Equation (8), resulted in a value of 0.56. This was used in Equation (6) to generate the fit curve shown in Figure 8. The compressibility index of the fit curve was 0.55 ± 0.04, and the resistance coefficient was $3.33 \cdot 10^{14} \pm 0.06 \cdot 10^{14}$ $m^2 \cdot s^2 \cdot kg^{-2}$. This is comparable to the compressibility indices of 0.5–1.7 reported for *E. coli*, depending on the membrane type and ionic strength [38]. Furthermore, the compressibility indices of rod-shaped *Bacillus circulans* (n = 1.0), *Rhodpseudomonas spheroides* (n = 0.88), and *E. coli* (n = 0.79) were compared with the ellipsoid cells of *Saccharomyces cerevisiae* (n = 0.45) and *Micrococcus glutamicus* (n = 0.31) during dead-end filtration [39]. The compressibility index of *V. natriegens* falls between these two sets of values, perhaps because cross-flow filtration produces a different cake structure based on the oriented stacking of rod-like cells rather than random packing [28]. For $n < 1$, the flux can be increased by increasing pressure; however, the cake resistance increases faster than increasing TMP when $n > 1$, leading to a critical flux at a certain TMP. Further increasing TMP does not gain more flux. This is critical to the process design and selection of process parameters.

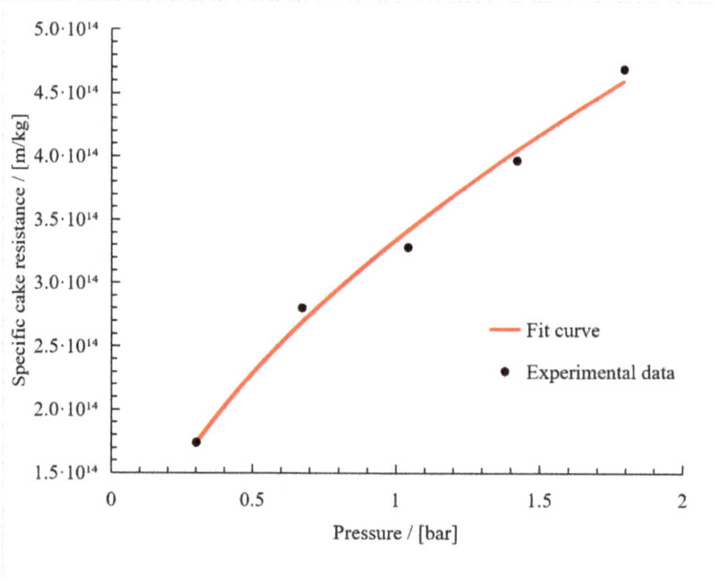

Figure 8. Fit of the specific cake resistance plotted against the pressure over the cake ($R^2 = 0.99$).

3.3. *Application of Hermia's Laws*

The permeate curves of each filtration were fitted using Hermia's laws for cross-flow filtration (Figure 9) and dead-end filtration (Figure 10). The resulting R^2 values are listed in Tables 3 and 4. The feeds were the culture broth, cell-free supernatant, cells in NaCl, and cultivation medium.

For the culture broth, the calculated R^2 values were contradictory. For the fitted formulas, the best model was the standard blocking model ($R^2 = 0.74$), but this was not supported by the appearance of the fit curve. Furthermore, only 14% of the resistance was reversible, which would rule out complete blocking and standard blocking as dominant fouling mechanisms according to Figure 1. The application of Hermia's non-adapted laws improved the fit for the cake filtration model ($R^2 = 0.88$). This is the highest value we observed and, therefore, represents the best-fitting model.

For the cell suspension, the best fit was the cake filtration model based on the adapted Hermia's laws ($R^2 = 0.92$), although the non-adapted laws also yielded a high value ($R^2 = 0.88$). This was supported by the appearance of the fit curves. The irreversible portion of fouling (29%) was also consistent with these conclusions.

For the supernatant, the calculated R^2 values were again contradictory. The reversible fouling share (89%) excluded standard and complete blocking as dominant fouling mechanisms. Nevertheless, standard blocking resulted in the best fit for the adapted Hermia's laws ($R^2 = 0.58$), but this was not consistent with the visual appearance of the fit curve. The equation with the best fit using the non-adapted laws was cake filtration ($R^2 = 0.71$).

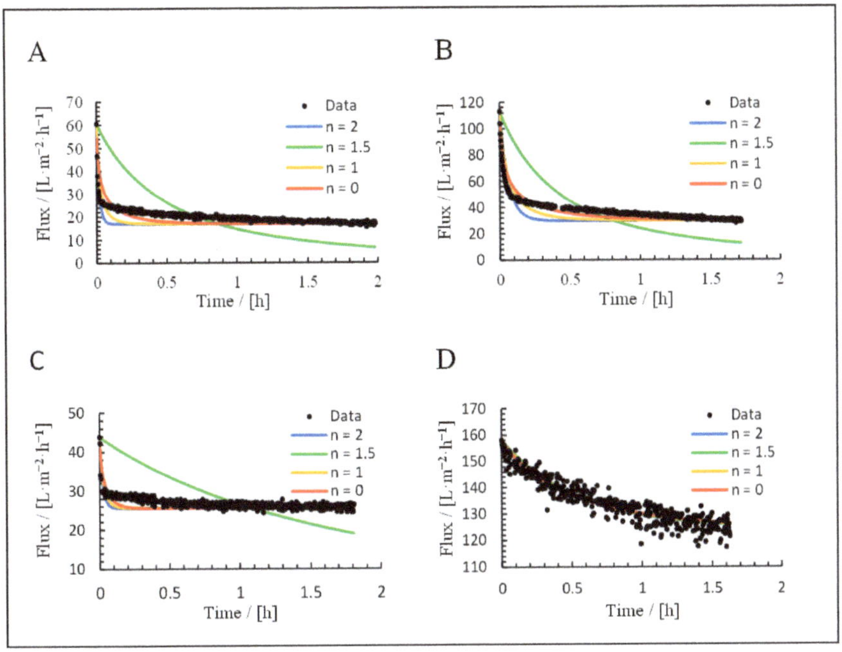

Figure 9. Filtration fit curves using Hermia's adapted laws: (**A**) culture broth, (**B**) = cells, (**C**) = supernatant, and (**D**) = medium.

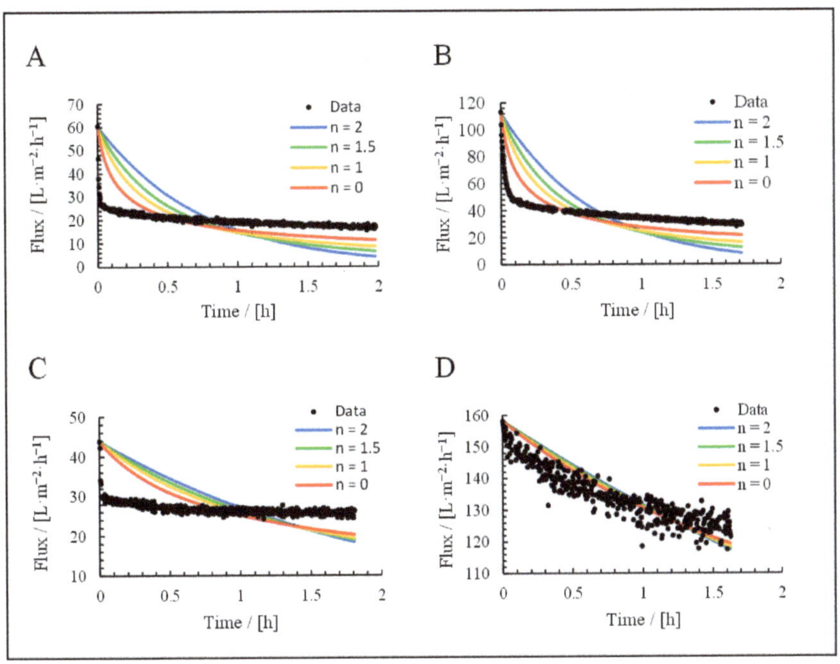

Figure 10. Filtration fit curves using Hermia's non-adapted laws: (**A**) culture broth, (**B**) = cells, (**C**) = supernatant, and (**D**) = medium.

Table 3. R^2 values representing fits of Hermia's laws adapted for cross-flow filtration (the highest values, implicating the best of the fits, are underlined).

Fit Model	R^2 of the Fit			
	Culture Broth	**Cells**	**Supernatant**	**Medium**
Complete blocking	0.53	0.81	0.50	<u>0.88</u>
Standard blocking	<u>0.74</u>	0.74	<u>0.58</u>	<u>0.88</u>
Intermediate blocking	0.61	0.88	0.53	<u>0.88</u>
Cake filtration	0.72	<u>0.92</u>	0.51	<u>0.88</u>

Table 4. R^2 values representing fits of Hermia's laws for dead-end filtration (the highest values, implicating the best of the fits, are underlined).

Fit Model	R^2 of the Fit			
	Culture Broth	**Cells**	**Supernatant**	**Medium**
Complete blocking	0.71	0.69	0.56	0.86
Standard blocking	0.74	0.74	0.58	0.87
Intermediate blocking	0.31	0.14	0.30	0.87
Cake filtration	<u>0.88</u>	<u>0.88</u>	<u>0.71</u>	<u>0.88</u>

For the medium, the best fit was cake filtration based on the adapted Hermia's laws (R^2 = 0.88), although the non-adapted laws delivered similar results. This was supported by the visual appearance of the fit curve and the reversible fouling share of 92%. Based on a visual comparison, the adapted formulas, therefore, provide a more suitable fit in all applications.

4. Conclusions

The results of the resistance in series model were conclusive. The individual resistances of the culture components were comparable to those of the culture broth. The instability of the cells in NaCl is an important finding that must be taken into account for future applications, and it should lead to process adaptations (especially shorter filtration times). The method developed to determine the compressibility index during cross-flow filtration is simple and will allow the compressibility index to be determined for other organisms with little effort. We found that the filtration time and feed composition were important parameters that must be investigated to ensure reliable results. The data will be useful to predict the filtration behavior of *V. natriegens*. However, factors such as a change in overflow velocity could limit the accuracy of such predictions and should be investigated in more detail. The application of Hermia's laws revealed that there is no dominant fouling mechanism during the filtration of the culture broth. This was anticipated because the culture broth is a complex mixture of cells, proteins, and salts. Even the separation of the culture broth into different components did not lead to clear results other than in the case of the cell suspension, where cake filtration was the dominant mechanism. Our data also show that the models have limited applicability when using biological feed streams. Factors such as cell lysis, metabolism, and associated temporal changes in feed composition require more complex models. Nevertheless, we have provided a starting point for the analysis of fouling during the cross-flow filtration of *V. natriegens* cultures, expanding our knowledge of the specific challenges and potential solutions.

Author Contributions: Conceptualization, P.C.; methodology, S.S. and R.F.; formal analysis, S.S.; investigation, S.S.; resources, P.C. and M.E.; writing—original draft preparation, S.S.; writing—review and editing, R.F., M.E. and P.C.; visualization, S.S.; supervision, P.C.; project administration, P.C.; funding acquisition, P.C. All authors have read and agreed to the published version of the manuscript.

Funding: This work was financially supported by the Hessian Ministry of Education and Art within the Hessian Initiative for Supporting Economic and Scientific Excellence (LOEWE) in the LOEWE-Research Center "DRUID—Novel Drug Targets against Poverty-Related and Neglected Tropical Infectious Diseases".

Institutional Review Board Statement: Not applicable.

Data Availability Statement: The original contributions presented in this study are included in this article. Further inquiries can be directed to the corresponding author.

Acknowledgments: We would like to thank Oliver Birrenbach for help with the programming of Labvision, Lisa-Marie Sittek for the friendly introduction to the use of the MCR102 modular compact rheometer, and Richard M. Twyman for professional language editing.

Conflicts of Interest: The authors declare no conflict of interest.

References

1. Austin, B.; Zachary, A.; Colwell, R.R. Recognition of Beneckea natriegens (Payne et al.) Baumann et al. as a member of the genus vibrio, as previously proposed by Webb and Payne. *Int. J. Syst. Bacteriol.* **1978**, *28*, 315–317. [CrossRef]
2. Payne, W.J. Studies on bacterial utilization of uronic acids. III. Induction of oxidative enzymes in a marine isolate. *J. Bacteriol.* **1958**, *76*, 301–307. [CrossRef] [PubMed]
3. Xu, J.; Dong, F.; Wu, M.; Tao, R.; Yang, J.; Wu, M.; Jiang, Y.; Yang, S.; Yang, L. Vibrio natriegens as a pET-Compatible Expression Host Complementary to Escherichia coli. *Front. Microbiol.* **2021**, *12*, 627181. [CrossRef]
4. Eagon, R.G. Pseudomonas natriegens, a marine bacterium with a generation time of less than 10 minutes. *J. Bacteriol.* **1962**, *83*, 736–737. [CrossRef]
5. Maida, I.; Bosi, E.; Perrin, E.; Papaleo, M.C.; Orlandini, V.; Fondi, M.; Fani, R.; Wiegel, J.; Bianconi, G.; Canganella, F. Draft genome sequence of the fast-growing bacterium Vibrio natriegens strain DSMZ 759. *Genome Announc.* **2013**, *1*, e00648-13. [CrossRef]
6. Hoffart, E.; Grenz, S.; Lange, J.; Nitschel, R.; Müller, F.; Schwentner, A.; Feith, A.; Lenfers-Lücker, M.; Takors, R.; Blombach, B. High substrate uptake rates empower Vibrio natriegens as production host for industrial biotechnology. *Appl. Environ. Microbiol.* **2017**, *83*, e01614-17. [CrossRef]
7. Schwarz, S.; Gerlach, D.; Fan, R.; Czermak, P. GbpA as a secretion and affinity purification tag for an antimicrobial peptide produced in Vibrio natriegens. *Electron. J. Biotechnol.* **2022**, *56*, 75–83. [CrossRef]
8. Díaz, E.; Blázquez, M.L.; Fernández-Llamosas, H.; Carmona, M.; Castro, L. Speeding up bioproduction of selenium nanoparticles by using Vibrio natriegens as microbial factory. *Sci. Rep.* **2017**, *7*, 16046. [CrossRef]
9. Eichmann, J.; Oberpaul, M.; Weidner, T.; Gerlach, D.; Czermak, P. Selection of High Producers From Combinatorial Libraries for the Production of Recombinant Proteins in Escherichia coli and Vibrio natriegens. *Front. Bioeng. Biotechnol.* **2019**, *7*, 254. [CrossRef]
10. Becker, W.; Wimberger, F.; Zangger, K. Vibrio natriegens: An Alternative Expression System for the High-Yield Production of Isotopically Labeled Proteins. *Biochemistry* **2019**, *58*, 2799–2803. [CrossRef]
11. Schleicher, L.; Muras, V.; Claussen, B.; Pfannstiel, J.; Blombach, B.; Dibrov, P.; Fritz, G.; Steuber, J. Vibrio natriegens as host for expression of multisubunit membrane protein complexes. *Front. Microbiol.* **2018**, *9*, 2537. [CrossRef] [PubMed]
12. Weinstock, M.T.; Hesek, E.D.; Wilson, C.M.; Gibson, D.G. Vibrio natriegens as a fast-growing host for molecular biology. *Nat. Methods* **2016**, *13*, 849–851. [CrossRef] [PubMed]
13. Des Soye, B.J.; Davidson, S.R.; Weinstock, M.T.; Gibson, D.G.; Jewett, M.C. Establishing a High-Yielding Cell-Free Protein Synthesis Platform Derived from Vibrio natriegens. *ACS Synth. Biol.* **2018**, *7*, 2245–2255. [CrossRef] [PubMed]
14. Lee, H.H.; Ostrov, N.; Wong, B.G.; Gold, M.A.; Khalil, A.S.; Church, G.M. Functional genomics of the rapidly replicating bacterium Vibrio natriegens by CRISPRi. *Nat. Microbiol.* **2019**, *4*, 1105–1113. [CrossRef]
15. Stadler, K.A.; Becker, W.; Darnhofer, B.; Birner-Gruenberger, R.; Zangger, K. Overexpression of recombinant proteins containing non-canonical amino acids in Vibrio natriegens: P-azido-L-phenylalanine as coupling site for 19F-tags. *Amino Acids* **2022**, *54*, 1041–1053. [CrossRef]

16. Jørgensen, C.E.; Abrahamsen, R.K.; Rukke, E.O.; Johansen, A.G.; Schüller, R.B.; Skeie, S.B. Optimization of protein fractionation by skim milk microfiltration: Choice of ceramic membrane pore size and filtration temperature. *J. Dairy Sci.* **2016**, *99*, 6164–6179. [CrossRef]
17. Cimini, A.; Moresi, M. Beer Clarification by Novel Ceramic Hollow-Fiber Membranes: Effect of Pore Size on Product Quality. *J. Food Sci.* **2016**, *81*, E2521–E2528. [CrossRef]
18. Moreno-García, J.; García-Martínez, T.; Mauricio, J.C.; Moreno, J. Yeast immobilization systems for alcoholic wine fermentations: Actual trends and future perspectives. *Front. Microbiol.* **2018**, *9*, 241. [CrossRef]
19. Oppermann, T. Mannheimia haemolytica—Bioprocess improvement of leukotoxin production using a fed-batch process and a ceramic membrane based downstream strategy. Ph.D. Thesis, Justus-Liebig-Universität Gießen, Giessen, Germany, 2018.
20. Ebrahimi, M.; Busse, N.; Kerker, S.; Schmitz, O.; Hilpert, M.; Czermak, P. Treatment of the bleaching effluent from sulfite pulp production by ceramic membrane filtration. *Membranes* **2015**, *6*, 7. [CrossRef]
21. Tomczak, W.; Gryta, M. Clarification of 1,3-propanediol fermentation broths by using a ceramic fine UF membrane. *Membranes* **2020**, *10*, 319. [CrossRef]
22. Persson, A.; Jönsson, A.S.; Zacchi, G. Separation of lactic acid-producing bacteria from fermentation broth using a ceramic microfiltration membrane with constant permeate flow. *Biotechnol. Bioeng.* **2001**, *72*, 269–277. [CrossRef] [PubMed]
23. Ebrahimi, M.; Kerker, S.; Daume, S.; Geile, M.; Ehlen, F.; Unger, I.; Schütz, S.; Czermak, P. Innovative ceramic hollow fiber membranes for recycling/reuse of oilfield produced water. *Desalin. Water Treat.* **2015**, *55*, 3554–3567. [CrossRef]
24. Hoffmann, D.; Leber, J.; Loewe, D.; Lothert, K.; Oppermann, T.; Zitzmann, J.; Weidner, T.; Salzig, D.; Wolff, M.; Czermak, P. *Purification of New Biologicals Using Membrane-Based Processes*; Elsevier Inc.: Amsterdam, The Netherlands, 2018. [CrossRef]
25. Maguire, N.A.P.; Ebrahimi, M.; Fan, R.; Gießelmann, S.; Ehlen, F.; Schütz, S.; Czermak, P. Influence of Ceramic Membrane Surface Characteristics on the Flux Behavior of a Complex Fermentation Broth. *Membranes* **2021**, *11*, 402. [CrossRef] [PubMed]
26. McCarthy, A.A.; Walsh, P.K.; Foley, G. Experimental techniques for quantifying the cake mass, the cake and membrane resistances and the specific cake resistance during crossflow filtration of microbial suspensions. *J. Memb. Sci.* **2002**, *201*, 31–45. [CrossRef]
27. Di Bella, G.; Di Trapani, D. A brief review on the resistance-in-series model in membrane bioreactors (MBRs). *Membranes* **2019**, *9*, 24. [CrossRef]
28. Mota, M.; Teixeira, J.A.; Yelshin, A. Influence of cell-shape on the cake resistance in dead-end and cross-flow filtrations. *Sep. Purif. Technol.* **2002**, *27*, 137–144. [CrossRef]
29. Riesmeier, B.; Kroner, K.H.; Kula, M.R. Studies on secondary layer formation and its characterization during cross-flow filtration of microbial cells. *J. Memb. Sci.* **1987**, *34*, 245–266. [CrossRef]
30. Tanaka, T.; Tsuneyoshi, S.I.; Kitazawa, W.; Nakanishi, K. Characteristics in Crossflow Filtration Using Different Yeast Suspensions. *Sep. Sci. Technol.* **1997**, *32*, 1885–1898. [CrossRef]
31. Vyas, H.K.; Mawson, A.J.; Bennett, R.J.; Marshall, A.D. A new method for estimating cake height and porosity during crossflow filtration of particulate suspensions. *J. Memb. Sci.* **2000**, *176*, 113–119. [CrossRef]
32. Diez, V.; Ezquerra, D.; Cabezas, J.L.; García, A.; Ramos, C. A modified method for evaluation of critical flux, fouling rate and in situ determination of resistance and compressibility in MBR under different fouling conditions. *J. Memb. Sci.* **2014**, *453*, 1–11. [CrossRef]
33. Gul, A.; Hruza, J.; Yalcinkaya, F. Fouling and chemical cleaning of microfiltration membranes: A mini-review. *Polymers* **2021**, *13*, 846. [CrossRef] [PubMed]
34. Shi, X.; Tal, G.; Hankins, N.P.; Gitis, V. Fouling and cleaning of ultrafiltration membranes: A review. *J. Water Process Eng.* **2014**, *1*, 121–138. [CrossRef]
35. Slimane, F.Z.; Ellouze, F.; Amar, N. Ben Fouling mechanism and screening of backwash parameters: Seawater ultrafiltration case. *Environ. Eng. Res.* **2019**, *24*, 298–308. [CrossRef]
36. Liang, J.; Yu, L.; Wu, J. The dynamic change of specific cake resistance in membrane bioreactor due to periodical cake relaxation. *J. Environ. Chem. Eng.* **2020**, *8*, 103837. [CrossRef]
37. Ghaffour, N.; Qamar, A. Membrane fouling quantification by specific cake resistance and flux enhancement using helical cleaners. *Sep. Purif. Technol.* **2020**, *239*, 116587. [CrossRef]
38. Fane, A.G.; Fell, C.J.D.; Hodgson, P.H.; Leslie, G.; Marshall, K. Microfiltration of biomass and biofluids: Effects of membrane morphology and operating conditions. *Filtr. Sep.* **1991**, *28*, 332–340. [CrossRef]
39. Nakanish, K.; Tadokoro, T.; Matsuno, R. On the Specific Resistance of Cakes of Microorganisms. *Chem. Eng. Commun.* **1987**, *62*, 187–201. [CrossRef]
40. Tanaka, T.; Abe, K.I.; Asakawa, H.; Yoshida, H.; Nakanishi, K. Filtration characteristics and structure of cake in crossflow filtration of bacterial suspension. *J. Ferment. Bioeng.* **1994**, *78*, 455–461. [CrossRef]
41. Loginov, M.; Doudiès, F.; Hengl, N.; Pignon, F.; Gésan-Guiziou, G. Influence of membrane resistance on swelling and removal of colloidal filter cake after filtration pressure release. *J. Memb. Sci.* **2020**, *595*, 117498. [CrossRef]
42. Sørensen, B.L.; Sørensen, P.B. Structure Compression in Cake Filtration. *J. Environ. Eng.* **1997**, *123*, 345–353. [CrossRef]

43. Kujundzic, E.; Greenberg, A.R.; Fong, R.; Moore, B.; Kujundzic, D.; Hernandez, M. Biofouling potential of industrial fermentation broth components during microfiltration. *J. Memb. Sci.* **2010**, *349*, 44–55. [CrossRef]
44. Fonseca, A.C.; Summers, R.S.; Greenberg, A.R.; Hernandez, M.T. Extra-cellular polysaccharides, soluble microbial products, and natural organic matter impact on nanofiltration membranes flux decline. *Environ. Sci. Technol.* **2007**, *41*, 2491–2497. [CrossRef] [PubMed]
45. Rushton, A. (Ed.) *Mathematical Models and Design Methods in Solid-Liquid Separation*; Springer: Dordrecht, The Netherlands, 1985. [CrossRef]
46. Bowen, W.R.; Calvo, J.I.; Hernández, A. Steps of membrane blocking in flux decline during protein microfiltration. *J. Memb. Sci.* **1995**, *101*, 153–165. [CrossRef]
47. Burghardt, J.P.; Coletta, L.A.; van der Bolt, R.; Ebrahimi, M.; Gerlach, D.; Czermak, P. Development and characterization of an enzyme membrane reactor for fructo-oligosaccharide production. *Membranes* **2019**, *9*, 148. [CrossRef]
48. Field, R.W.; Wu, D.; Howell, J.A.; Gupta, B.B. Critical flux concept for microfiltration fouling. *J. Memb. Sci.* **1995**, *100*, 259–272. [CrossRef]
49. Gunde-Cimerman, N.; Plemenitaš, A.; Oren, A. Strategies of adaptation of microorganisms of the three domains of life to high salt concentrations. *FEMS Microbiol. Rev.* **2018**, *42*, 353–375. [CrossRef]
50. Gregory, G.J.; Boyd, E.F. Stressed out: Bacterial response to high salinity using compatible solute biosynthesis and uptake systems, lessons from Vibrionaceae. *Comput. Struct. Biotechnol. J.* **2021**, *19*, 1014–1027. [CrossRef]
51. Magpusao, J.; Oey, I.; Kebede, B. Opportunities and Challenges of Algal Protein Extraction and Production. *Innov. Food Process. Technol. A Compr. Rev.* **2020**, *3*, 216–233. [CrossRef]
52. Liu, D.; Wu, C.; Wu, R.; Huang, J.; Liao, B.; Lei, M.; Zhang, Y.; He, H. Comprehensive analysis of the phylogeny and extracellular proteases in genus Vibrio strain. *Microb. Pathog.* **2019**, *131*, 1–8. [CrossRef]
53. Thoma, F.; Blombach, B. Metabolic engineering of Vibrio natriegens. *Essays Biochem.* **2021**, *65*, 381–392. [CrossRef]

membranes

Article

Enhancing the Performance of Tangential Flow Microfiltration for Bioreactor Clarification

Amir Hossein Mostafavi [1], Liang-Kai Chu [2,†], Xianghong Qian [3], John Paul Smelko [4], Da Zhang [1,‡], Andrew Zydney [2] and Sumith Ranil Wickramasinghe [1,*]

1 Ralph E. Martin Department of Chemical Engineering, University of Arkansas, Fayetteville, AR 72701, USA; am223@uark.edu (A.H.M.); dzhang@benitec.com (D.Z.)
2 Department of Chemical Engineering, The Pennsylvania State University, University Park, PA 16802, USA; liang-kai.chu@merck.com (L.-K.C.); zydney@engr.psu.edu (A.Z.)
3 Department of Biomedical Engineering, University of Arkansas, Fayetteville, AR 72701, USA; xqian@uark.edu
4 Biogen, 5000 Davis Drive, Research Triangle Park, NC 27709, USA; john.smelko@biogen.com
* Correspondence: ranil.wickramasinghe@uark.edu
† Current address: Merck & Co., Inc., West Point, PA 19486, USA.
‡ Current address: Benitec Biopharma Inc., 3940 Trust Way, Hayward, CA 94545, USA.

Abstract: Tangential flow microfiltration is easily adapted for batch and continuous bioreactor clarification. The permeate can be introduced directly to the subsequent capture step. However, the commercial use of tangential flow filtration (TFF) is limited by membrane fouling, leading to a compromised performance. Here, we explored the possibility of reducing membrane fouling by integrating a hydrocyclone as the primary clarification operation. The overflow from the hydrocyclone was introduced directly as the feed to the microfiltration module. Chinese hamster ovary cells were used as the feed stream to investigate the feasibility of this integrated process. A range of cell viabilities from 0% (cell lysate) to 96% were investigated. The cell densities ranged from 0.9 to 10 million cells per mL. Two commercially available hollow fiber microfiltration membranes were used, an essentially symmetric membrane and a reverse asymmetric membrane where the more open support structure faced the feed stream. The reverse asymmetric membrane was more resistant to fouling in the absence of an integrated hydrocyclone. Integrating a hydrocyclone led to a reduction in the flux decline for the symmetric membrane, but did not affect the performance of the reverse asymmetric membrane. The careful choice of membrane morphology and pore size is important when designing an integrated process.

Keywords: bioreactor harvesting; clarification; continuous biomanufacturing; fouling; hydrocyclone; membrane morphology; microfiltration

1. Introduction

Biotechnology manufacturing processes make use of cell culture operations to grow cells that produce the desired therapeutic (e.g., monoclonal antibodies, virus vectors, etc.). Bioreactor clarification operations are used to remove the unwanted particulate matter (cells, cell debris, etc.) from the growth medium that contains the product of interest. Typically, manufacturing processes are run in batch mode.

However, there is significant interest in the development of continuous biomanufacturing processes [1]. The major advantages of continuous manufacturing processes are lower capital and operating costs. This results from the more efficient utilization of the process equipment, a smaller footprint and equipment size, and a higher productivity due to less equipment downtime [2]. However, in the biopharmaceutical industry, the cost of goods has not been the dominant driver, explaining the slower development of continuous biomanufacturing processes. This has begun to change with the development of biosimilars and the desire to enable greater US-based manufacturing to reduce the reliance on overseas supply chains. Additionally, continuous biomanufacturing can lead to improved product quality resulting from the much shorter residence times and steady state operation.

Most of the advances in continuous biomanufacturing have focused on upstream processes [3]. Significant progress has been made on the development of continuous cell culture processes, with media added continuously while the product is removed in what is commonly referred to as a perfusion bioreactor. Cost analyses indicate that higher cell densities obtained in continuous cell culture operations lead to higher production rates and lower costs [4]. Unlike batch operations, where the bioreactor is clarified at the end of the cell culture operation, the product needs to be removed continuously while retaining the cells in perfusion operations. Consequently, some type of cell retention device must be integrated with the bioreactor. A variety of cell retention devices have been explored with different methods of action, such as filtration, sedimentation, centrifugation, ultrasonic fixation, and dielectrophoretic exclusion [5]. Filtration-based methods are of particular interest since they can effectively remove both whole cells and relatively small cell debris, depending upon the pore size of the filter. Here, we focused on tangential flow filtration (TFF), as it can easily be incorporated into batch and continuous manufacturing processes.

TFF has many advantages over other bioreactor clarification operations for both batch and continuous processing [6]. If an appropriate-pore-size membrane is used, e.g., 0.22 or 0.45 mm, the permeate is sufficiently "clean" to be directly fed to the subsequent capture step (most commonly Protein A chromatography for monoclonal antibody production). However, the high cell density in state-of-the-art bioreactors often causes membrane fouling, leading to product rejection and a low filtrate flux, compromising the viability of TFF. Further, especially for emerging therapeutics such as virus vectors, cell lysis is often required, as the product of interest is not excreted into the growth medium. TFF for the clarification of cell lysates is extremely challenging due to rapid membrane fouling.

TFF can easily be incorporated as a cell retention device in perfusion operations. However, high cell densities tend to lead to higher rates of fouling. Higher rates of fouling invariably lead to product sieving losses during operation, especially over extended periods of time. Three methods may be used to minimize fouling during TFF. The membrane structure can be modified. Da et al. [7,8] indicated that the use of a reverse asymmetric membrane where the more open pore structure faces the feed stream and the barrier layer faces the permeate can lead to higher productivity during batch operation. Alternatively, one can modify the TFF operating conditions. For example, alternating tangential filtration (ATF), in which the direction of the feed flow is alternated in a cyclic fashion, can minimize fouling, but the operating filtrate flux in these systems is typically <3 L m^{-2} h^{-1} [9]; thus, large membrane areas are required for commercial manufacturing operations.

A third approach for enhancing the performance of TFF that is appropriate for both batch and continuous operations is to integrate TFF as a secondary clarification step with another unit operation used to reduce the load of particulate matter (cells and cell debris) before performing TFF. In perfusion processes, the primary clarification operation must be amenable for continuous operation with the recycling of the retained cells. Thus, depth filtration, which is widely used for initial clarification, is not viable in this application. The incorporation of an appropriate continuous primary clarification step would reduce the particulate load in the feed to the TFF step, minimizing fouling and the associated reductions in product recovery during extended operation of the perfusion bioreactor. Modular chemical process intensification, where two unit operations are combined to achieve an integrated system with a higher performance than each of the individual unit operations, is of growing interest in the biotechnology industry [10,11]. Here, we investigated for the first time the use of a hydrocyclone as the primary clarification operation in combination with TFF as the secondary clarification operation.

Hydrocyclones have been used for many years for particle separations from liquid streams in the mineral-processing, chemical, and petrochemical industries. They can be easily scaled up and used in continuous manufacturing and have low operating costs [12,13]. The feed stream is introduced at the top of the hydrocyclone, as shown in Figure 1a. There are two exits from the hydrocyclone: the underflow, which contains the larger (higher-density) particulate matter, and the overflow, which contains a low concentration of smaller (lower-density) particles. The larger and higher-density particles move towards the wall due to centrifugal and drag forces as well as the pressure gradient. They then move downwards and into the underflow. The smaller and lower-density particles move upwards and exit with the overflow [14]. The degree of separation of the particles into the underflow depends on the rate of centrifugal sedimentation. While the principle of separation is the same as in classical centrifugation, the hydrocyclone has no moving parts, eliminating many of the challenges of large-scale continuous centrifuges.

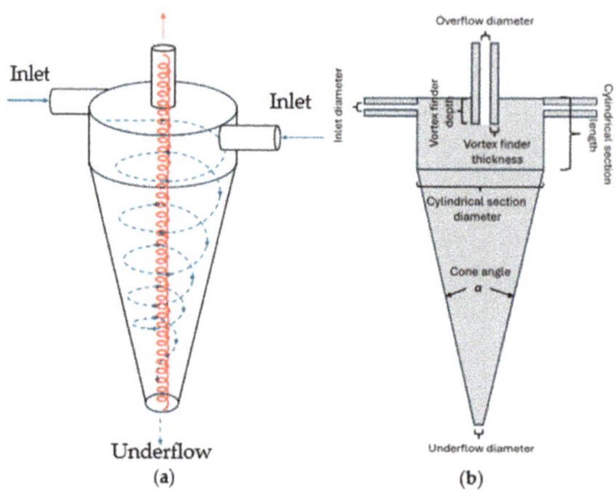

Figure 1. Schematic of a hydrocyclone; (**a**) flow pattern and (**b**) nomenclature for relevant components.

Several investigators have examined the use of a hydrocyclone alone for mammalian cell separations [15,16]. These studies indicate that, for CHO cells, a separation efficiency in the overflow versus underflow as high as 99% can be obtained with cell viabilities as high as 90% at pressure drops of up to 3 bar. Hydrocyclones have also been used as a cell retention device for perfusion bioreactors where similar cell viabilities were achieved. Ni et al. [13] provided a review of the important design parameters.

Hydrocyclones are ideally suited for the separation of particles 10 mm or larger, but they are much less effective at removing smaller cell debris. Particles of a given size will reach an equilibrium radial position where the outward terminal settling velocity equals the inward radial velocity of the liquid (see Figure 1a). Particles with an equilibrium position that coincides with the locus of zero liquid vertical velocity have an equal probability of leaving via the underflow or overflow. This is known as the cut size. Reducing the cut size by optimizing the various geometric parameters will lead to higher feed pressures. However, high feed pressures are undesirable for perfusion bioreactors, as they will adversely affect cell viability. Thus, a secondary clarification operation will be required after the hydrocyclone to obtain a product stream suitable for subsequent downstream processing operations.

Syed et al. [17] used a hydrocyclone for primary cell separation, with a spiral microchannel used for secondary clarification, as the cell retention device for a perfusion bioreactor. They indicated up to 75% separation of cells by the hydrocyclone with <5% loss in cell viability. The spiral microchannel system removed more than 90% of the remaining particulate matter; thus, some type of additional clarification would likely still be needed. In addition, scaling up the spiral microchannel device is likely to be challenging. He et al. [18] investigated in detail the viability loss of CHO cells for a continuous cell culture using hydrocyclones. They indicate that cell viability decreases with an increase in the hydrocyclone cone angle (see Figure 1b) and inlet velocity due to the increased pressure drop to which the cells are subjected. The highest separation efficiencies obtained by He et al. [18] were <90%, although there was no discussion of how the required secondary clarification would be performed.

The current study provides the first analysis of using a hydrocyclone in combination with TFF in order to improve the performance of the TFF operation. The hydrocyclone reduces the load of particulate matter in the feed stream to the microfiltration unit, significantly reducing the rate of fouling and increasing the performance of the TFF unit. Two different hydrocyclones were fabricated using 3D printing based on literature designs. Two different TFF modules were investigated, with one containing an essentially symmetric membrane and one having a reverse asymmetric membrane. The effects of the different membrane morphologies are also discussed. The results provide insights into the development of an integrated hydrocyclone–TFF bioreactor clarification process with applications in batch and continuous bioprocessing operations.

2. Materials and Methods

2.1. Materials

The reagents used in this study were biotechnology grade or higher if not specified. FreeStyle™ CHO-S cells were obtained from ThermoFisher (Waltham, MA, USA). CHOgro® Expression Media and the poloxamer 188 (Pluronic) solution, 10% w/v in cell-culture-grade water, were purchased from Mirus Bio (Madison, WI, USA). L-glutamine, TRITON™ X-100, and sodium hydroxide were purchased from VWR (Radnor, PA, USA). Trypan blue was purchased from Sigma-Aldrich (Burlington, MA, USA).

All the rubber tubing and peristaltic pumps used to deliver fluids were purchased from Masterflex (Vernon Hills, IL, USA). The water used in this work was obtained from a ThermoFisher 18 MΩ Barnstead Smart2pure system (Schwerte, Germany). An electronic scale (PL602-S) was purchased from Mettler Toledo (Columbus, OH, USA). Two hollow fiber microfiltration modules were custom manufactured for this work by Asahi Kasei Bioprocess (Glenview, Il, USA). The BioOptimal™ MF-SL filter had a membrane surface area of 0.00041 m^2, with a pore size of 0.4 μm on the permeate side and 40 μm on the feed side. The module is unique in that the more open support structure faces the feed, unlike conventional microfiltration modules in which the size-selective "skin" faces the feed. The UJP module has a membrane surface area of 0.00032 m^2. The barrier layer, which faces the feed stream, has a nominal pore size of 0.65 μm, as provided by the manufacturer. While there is variation in the average pore size from the inside to the outside of the fiber, the membrane is much more symmetric in structure. Both hollow fiber membranes were manufactured by a wet spinning process. The membrane materials were polyvinylidene-fluoride for the UJP and polysulfone for the BioOptimal MF-SL filters. The hydrocyclones were fabricated by 3D printing as described below.

2.2. Cell Culture

FreeStyle™ CHO-S cells were cultured using flasks in a shaker incubator at 37 °C and 125 rpm. A humidified atmosphere consisting of 8% CO_2 was used. Cell titers of up to 25 million cells/mL were obtained using CHOgro® Expression media supplemented with 8 mM L-glutamine and 0.3% pluronic. First, CHO cells were thawed and then passaged after a few days repeatedly to obtain the desired concentration. The cell density was evaluated under a microscope using a hemacytometer with trypan blue exclusion dye used to identify the number of dead cells. The trypan blue exclusion assay is based on the principle that live cells have intact cell membranes that will exclude the dye, while dead cells will turn blue. The cell sample was mixed with dye and then visually examined. Viable cells appeared clear while nonviable cells appeared blue.

The cell viability was determined by Equation (1):

$$Cell\ Viability\ (\%) = \frac{N_v}{N_t} \times 100 \tag{1}$$

where N_v and N_t are the number of viable and total cells, respectively.

2.3. Hydrocyclone Designs

There have been numerous studies that have considered the design of hydrocyclones [19]. Several geometrical and operational factors should be considered, among which the inlet and outlet diameters, cone angle, vortex finder depth and thickness, inlet pressure and velocity, and concentration of particulate matter are important [12,13]. The inlet and outlet diameters and ratios primarily influence the velocity of the fluid and the pressure drop. For mammalian cells, these two factors are important, as high pressures lead to a reduction in cell viability.

The hydrocyclones in this study were made using Fromlab 3-D printers (Somerville, MA, USA) using gray and clear resins (Figure 2). The designs were based on the hydrocyclones used in previous bioreactor clarification studies [17,18]. Table 1 shows the design parameters. Design I (larger hydrocyclone) is based on the work by He et al. [18], while design II is based on the work by Syed et al. [17]. As can be seen, design I is simpler and has the same inlet and outlet diameters. Both hydrocyclones are based on the Bradley hydrocyclone.

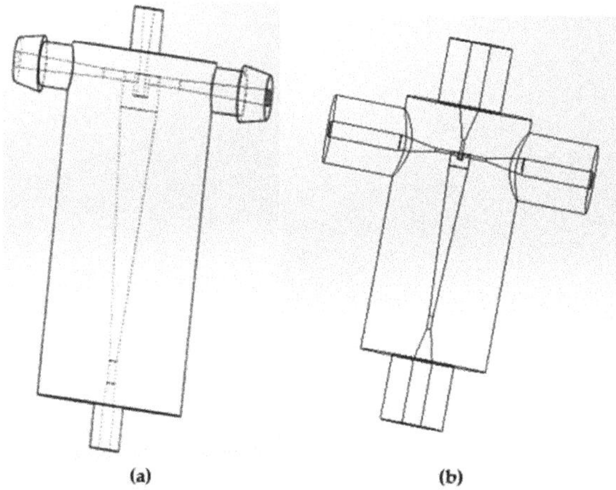

Figure 2. CAD drawings of hydrocyclones in SolidWorks. (**a**) Design I and (**b**) design II.

Table 1. Design parameters for hydrocyclones fabricated as part of this study. Design 1 is based on He et al. [18]. Design II is based on Syed et al. [17].

Parameter	Dimensions (mm) (Design I)	Dimensions (mm) (Design II)
Cylindrical section diameter	10	3
Overflow diameter	2	0.4
Underflow diameter	2	0.65
Inlet diameter	2	0.413
Length of the cylindrical section	8	2.16
Cone length	57.2	23.16
Vortex-finder depth	5	1
Vortex-finder thickness	1	0.2–0.4

Table 2 compares important design ratios for the two hydrocyclones to the ratios provided by Bradley [20]. The Bradley hydrocyclones have been shown to provide high separation efficiencies [19]. As can be seen, both hydrocyclones have elements of Bradley's optimized ratios. The SolidWorks PDM 2023 software (Cleverbridge, Inc., Chicago, IL, USA) was used to design the hydrocyclones prior to printing, as shown in Figure 2.

Table 2. Comparison of geometric ratios for the hydrocyclones in comparison with Bradley's ratios.

Ratio	Design I	Design II	Bradley's Ratio
Inlet/cylindrical section diameter	0.2	0.133	0.133
Overflow/cylindrical section diameter	0.2	0.133	0.2
Vortex finder depth/cylindrical section diameter	0.5	0.33	0.33
Cone length/cylindrical section diameter	5.7	7.7	6.58

2.4. Experimental Setup and Procedure

Three different clarification operations were considered: TFF only, hydrocyclone only, and an integrated hydrocyclone and TFF process, as shown in Figure 3.

Before starting each experiment, the hydrocyclone was washed with 10% bleach followed by DI water. Then, the supplemented media was circulated through the system prior to adding the cell suspension. The TFF filters were also cleaned, regenerated, and washed with PBS, a NaOH/bleach solution, and then DI water, in that order. For the hydrocyclones, prior to testing with the cell suspension, the underflow DI water split ratio was determined using Equation (2).

$$Split\ ratio\ (\%) = \frac{Q_U}{Q_F} \times 100 \tag{2}$$

where Q_U and Q_F are the underflow rate and the feed flow rate, respectively. Various cell densities were investigated. We compare our result to the work of He et al. [18] and Syed et al. [17]. He et al. determined the separation efficiency as

$$E\ (\%) = \left(\frac{Q_U C_U}{Q_U C_U + Q_O C_O}\right) \times 100 \tag{3}$$

where C_U and C_O are the cell concentrations in the underflow and overflow, respectively. Syed et al. defined the separation efficiency differently as

$$E\ (\%) = \left(1 - \frac{C_O}{C_F}\right) \times 100 \tag{4}$$

where C_F is the cell concentration in the feed. The results obtained here were compared to the respective literature data based on these definitions, given by Equations (3) and (4).

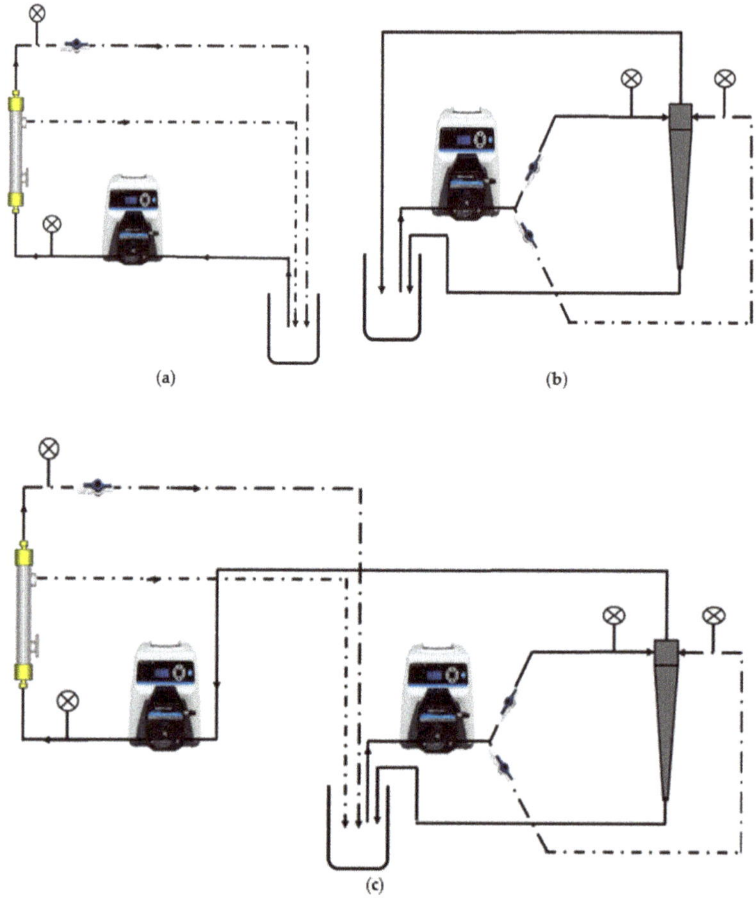

Figure 3. Experimental setups used here (**a**) TFF, (**b**) hydrocyclone, and (**c**) integrated hydrocyclone–TFF.

2.5. Bioreactor Harvesting

Prior to use, the microfiltration modules were flushed with 1 L of DI water. The DI water was pumped through the fiber lumen with the permeate line open, enabling water to pass through the retentate and permeate flow paths. The same procedure was used for the UJP and BioOptimal MF SL filters. This filter preparation was not expected to have any effect on filter performance.

The membrane was regenerated after use according to the following protocol. The filter was first flushed with 1–2 L of DI water at room temperature to remove cells and weakly bound debris. The feed was pumped through the fiber lumen and the permeate line was opened, allowing water to pass through the retentate and permeate lines. Next, the membrane was back-flushed (fluid pumped from outside the fibers through the membrane to the fiber lumen) using about 100 mL of 0.5 N NaOH with 3000 ppm NaClO at room temperature. The module was then forward-flushed (i.e., fluid flows from fiber lumen through the membrane to the shell side) using 0.5 N NaOH with 3000 ppm NaClO at 30 mL/min for 30 min at room temperature. Using the same solution composition, the module was back-flushed at 30 mL/min for 30 min. The last two steps were repeated

3–4 times. The filter was then stored overnight in 0.5 N NaOH with 3000 ppm NaClO. Finally, the next day, the filter was flushed (flow from the feed side to the retentate side) with 1–7 L of water until the pH was neutralized. The nominal water flux for the BioOptimal MF-SL was >21,000 L m^{-2} h^{-1}. For the UJP filter, the nominal water flux was 1200 L m^{-2} h^{-1}. Both fluxes were at a TMP of 0.1 MPa and a temperature of 25 °C.

Two sets of TFF experiments were run. The first series of experiments involved running the UJP and BioOptimal MF-SL filters at a wall shear rate of 2000 s^{-1} (feed flow rate of 16 mL min^{-1} for the UJP filter and 32 mL min^{-1} for the BioOptimal MF-SL filter) in concentration mode. A wall shear rate of 2000 s^{-1} was chosen based on the manufacturer's recommendations and previous studies. The experiments were run in batch concentration mode. The cell density was 9 million/mL and the feed volume was 50 mL. The initial transmembrane pressure (TMP) was 3.4 kPa (0.5 psi) and 13 kPa (2 psi) for the BioOptimal MF-SL and UJP filters, respectively. It is essential that the BioOptimal is run at a low TMP in order to prevent compaction of the stabilized cake layer within the more open membrane pore structure that faces the feed stream. These TMP values resulted in an initial permeate flux of 1100 L m^{-2} h^{-1} for both modules. Experiments were also run at different cell viabilities using the BioOptimal MF-SL filter under the same experimental conditions. The cell viability was decreased by reducing the rate of nutrient addition. The experiments were designed to investigate the effect of the reverse asymmetric membrane structure on flux decline.

The second set of TFF experiments was run with and without an integrated hydrocyclone. These experiments were run in total recycle mode. The feed flow rate to the hyrocyclone was 50 mL min^{-1}. The overflow from the hydrocyclone was introduced to the TFF module at a feed flow rate of 10 mL min^{-1}, corresponding to wall shear rates of 620 s^{-1} and 1280 s^{-1} for the BioOptimal MF-SL and UJP filters, respectively. The initial TMP for TFF only (without the hydrocyclone) was 3.7 $\pm$ 0.14 kPa (0.54 $\pm$ 0.02 psi) for the BioOptimal MF-SL and 12.4 $\pm$ 1.3 kPa (1.8 $\pm$ 0.19 psi) for the UJP filter. For the integrated system, the TMP was 4.0 $\pm$ 0.41 kPa (0.58 $\pm$ 0.06) and 12.1 $\pm$ 0.83 kPa (1.76 $\pm$ 0.12 psi) for the BioOptimal and UJP modules, respectively.

Additional experiments were conducted using the BioOptimal MF-SL filter in ATF mode. A Repligen (Waltham, MA, USA) alternating tangential flow-2 (ATF-2) filtration system was used for perfusion system setup. The setup used a diaphragm pump to both withdraw and return the cell suspension to the bioreactor. The perfusion rate was 300 mL min^{-1}, i.e., 300 mL of fluid was pumped in and out of the bioreactor in 1 min. Although no frequency was set, considering that the displacement of one exchange is 120 mL, about 20 s was estimated for 1 displacement (120 mL). The frequency was controlled by the ATF flow rate. The ATF flow rate is the most important parameter. The set points for the pH, temperature, and DO were 7, 37 °C, and 40%. A Sartorius (Göttingen, Germany) BioStat A plus with a starting cell density of 1 million viable cells/mL in a 1.5 L working volume was used. Perfusion was initiated one day after inoculation at 1 reactor volume per day (RV/day). The perfusion rate was then increased to 1.5 RV/day when the cell density approached a value of 10 million viable cells/mL. After that, the perfusion rate was maintained at 1.5 RV/day. The perfusion run was performed for 14 days.

3. Results

3.1. Membrane Structure

Figure 4 gives SEM images of the membranes in the UJP and BioOptimal MF-SL filters. These images were obtained from the manufacturer. As can be seen from Figure 4a, which gives a cross-sectional image of the UJP membrane, the membrane is rather symmetric. Figure 4b–e give various images of the BioOptimal MF-SL membrane. The membrane has a much more open pore structure in the fiber lumen which is in contact with the feed. The barrier layer is located at the outer fiber surface. As can be seen for the UJP membrane, the surface porosity appears similar though at the middle of the fiber the porosity is much less. In the case of the asymmetric BioOptimal membrane, the inner surface has a much higher porosity than the outer surface.

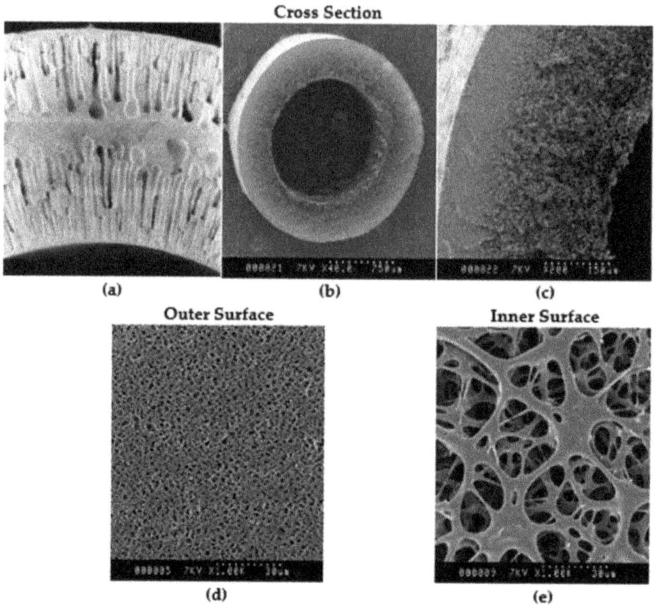

Figure 4. SEM images of (**a**) UJP cross section, with inside diameter of 1.1 mm, courtesy of Asahi Kasei Corporation; (**b–e**) BioOptimal MF-SL, with inside diameter of 1.4 mm; and (**b**,**c**) cross section showing clear asymmetric structure, (**d**) outside surface, and (**e**) inside surface, courtesy of Asahi Kasei Bioprocess.

In our earlier work [7,8], we showed that the more open structure stabilizes a highly permeable cake layer that acts as a depth filter. The stabilized cake layer removes particles that would otherwise foul the barrier layer.

3.2. *Effect of Cell Viability on TFF*

Figure 5 gives the permeate flux versus throughput (permeate recovered per membrane surface area) for the first set of TFF experiments. Data are included for the BioOptimal MF-SL and UJP modules at a wall shear rate of 2000 s^{-1} run in batch concentration mode. The cell viability was over 96% for the UJP. The lower-viability experiments for the BioOptimal MF-SL used cell suspensions that were partially lysed to reduce the cell viability. The initial TMP was chosen to give the same initial permeate flux, 1110 L m^{-2} h^{-1}. Since, industrially, TFF is run under constant flux conditions, the performance of the two modules was compared at the same flux. The results indicated a rapid flux decline for the UJP module as the contents of the feed reservoir became concentrated. The BioOptimal MF-SL displayed a much higher throughput during the batch concentration.

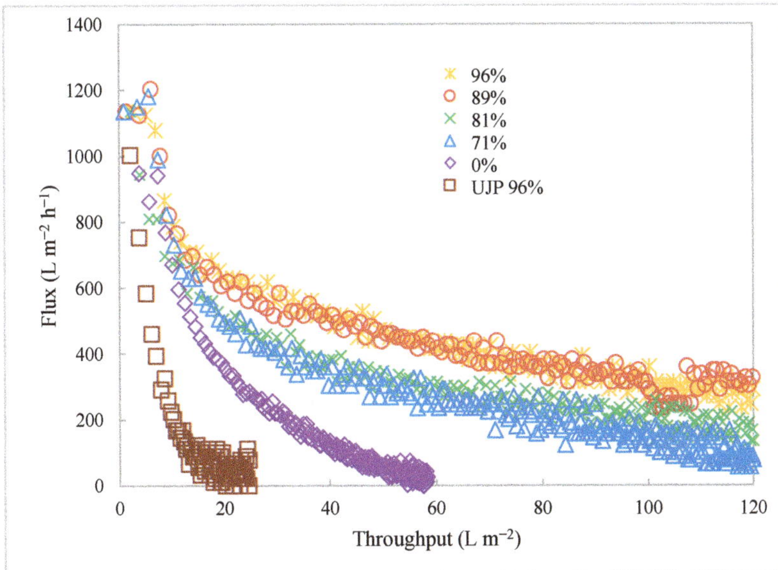

Figure 5. Flux versus productivity for BioOptimal MF-SL and UJP filters under a range of feed conditions. Percentages in the legend represent different cell viabilities. All results are for BioOptimal unless noted in the legend.

Figure 5 also gives the flux versus throughput results for the BioOptimal MF-SL for cell viabilities ranging from 0 to 89%. Given the very low throughput for the UJP membrane, even at cell viabilities of over 96%, it was not used to examine the effect of cell viability on the throughput. As can be seen, there was a rapid decrease in the throughput with reduced viability. When all the cells were dead, with 0% viability representative of a cell lysate, the throughput was at its lowest. Dead cells tend to lyse, leading to a large amount of small particulate matter that can easily deposit on and within the membrane. These results indicate that, although the reverse asymmetric structure of the BioOptimal MF-SL membrane can lead to much higher throughputs than observed for the UJP membrane, low-viability feed streams still cause extensive fouling, which would lead to significant challenges in the application of these membranes for cell recycle and primary clarification during long-term continuous perfusion operations.

3.3. Hydrocyclone Performance

Before using the hydrocyclones as part of the integrated clarification process, initial testing was conducted to determine if the performance of the hydrocyclones developed in this work were similar to the ones designed by He et al. [18] and Syed et al. [17]. Figure 6 gives the variation in the inlet velocity with the underflow DI water split ratio, as defined by Equation (2). The right-hand *y*-axis gives the pressure drop. As can be seen, for both designs, the percentage of the total feed that exits via the underflow (split ratio) was nearly independent of the inlet feed velocity over the range of conditions examined in Figure 6. However, the underflow split ratio was much higher for design II. The split ratio is affected by a number of factors. The important factors here are the overflow and underflow outlet ratios compared to the size of the inlet or cylindrical sections. Figure 1b and Table 1 give the important dimensions of the hydrocyclones. A larger underflow-to-inlet diameter ratio compared to the overflow-to inlet diameter ratio will lead to a higher split through the underflow. On the other hand, a larger vortex finder depth and a higher-pressure drop will lead to a higher underflow split ratio. Here, although design I had a larger vortex finder depth and displayed a higher pressure drop, the influence of the underflow-to-inlet diameter dominated, leading to the higher split ratio for design II. These results indicate that it was the combined effects of the relevant variables that led to the observed split ratio. The inlet pressure results also show that design I experienced a greater pressure increase with increasing velocity. The higher pressure drop was due to the steeper cone angle and lower outlet/inlet ratios.

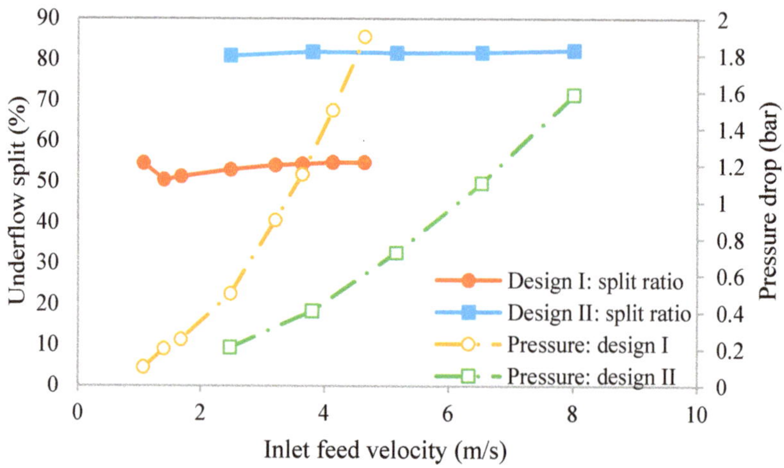

Figure 6. Variation in underflow split ratio and pressure drop with inlet feed velocity.

Figure 7 gives the separation efficiencies of the two hydrocyclones tested here. Figure 7a compares the efficiency of design I with the results obtained by He et al. [18], while Figure 7b compares the efficiency of design II with the results obtained by Syed et al. [17]. In both cases, the separation efficiency is plotted against the inlet feed velocity. Figure 7a,b indicate that, even though lower feed flow rates (and consequently, pressure drops) were used here, the results are in good agreement with those from previous studies using the same hydrocyclone designs. Over the range of cell densities investigated here, it appears that cell density has little effect on the separation efficiency. It is worth noting that the cell viability dropped significantly to less than 90% at higher operating velocities (similar to the literature studies), likely due to the very high pressures under these conditions. Therefore, our experiments focused on the performance at lower inlet velocities than reported in the literature. Specifically, for design I, the viability loss was more significant due to the higher normal stress (pressure), as shown in Figure 6.

Cell viability depends on the feed flow rate (shear stress to which the cells are subjected) and the time for which they are subjected to a given shear stress. With an increasing run time, the viability drops. However, since these runs were conducted under non-sterile conditions, the cell viability could also drop due to bacterial infection. For design 1, Figure 7a shows that, at all flow rates, the viability dropped over a 2 h period to below 90%. The decrease in viability was much greater at higher flow rates. For design II, only at the highest feed flow rate investigated did the viability drop to below 90% after 2 h.

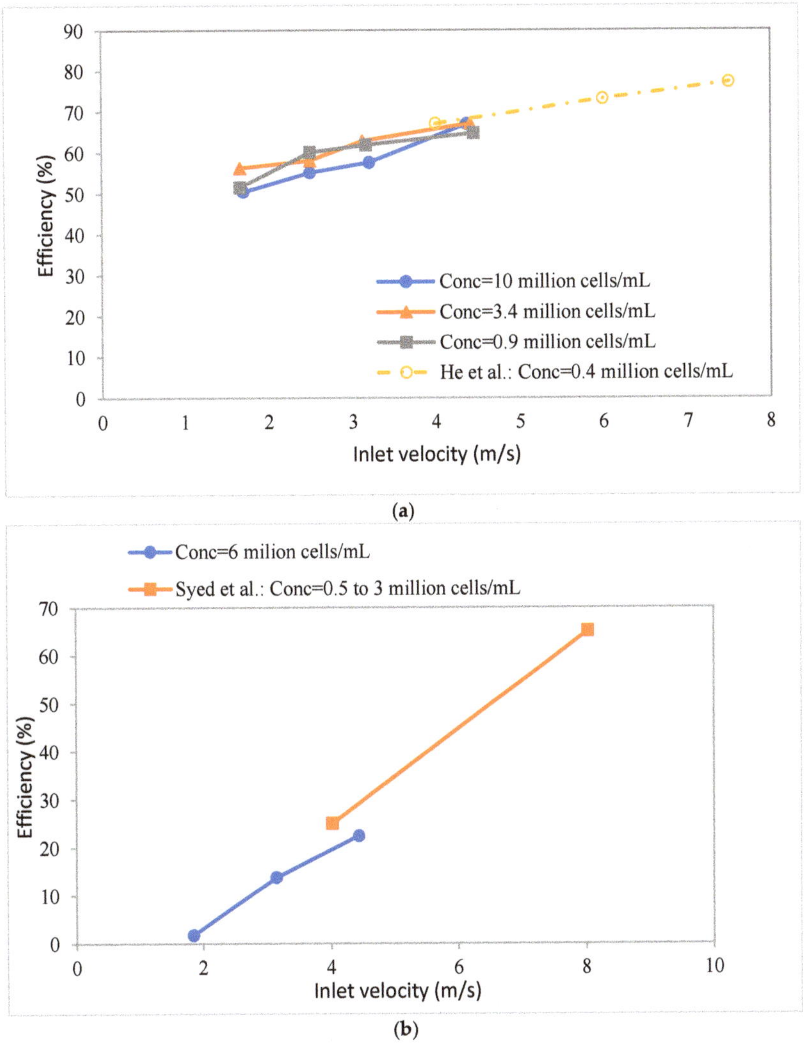

Figure 7. Separation efficiency versus inlet velocity. (**a**) Comparison of efficiency of hydrocyclone design 1 with results of He et al. [18]; (**b**) comparison of efficiency of design II with results of Syed et al. [17].

3.4. ATF with BioOptimal MF SL Microfilter

Figure 8 provides the results for the experiment conducted with the BioOptimal MF SL filter in ATF mode. The viable cell density (left-hand-side *y*-axis) and cell viability (right-hand-side *y*-axis) are plotted against the perfusion time in days. The run was conducted for 14 days. The perfusion rate was 1 vessel volume (1.5 L per day, or 1 mL/min) for the first 5 days (i.e., below 10 million viable cells/mL) and then 1.5 vessel volumes (2.25 L per day, or 1.5 mL/min) for the rest of the run (cell densities larger than 10 million viable cells/mL). The average permeate flux was about 152 L m^{-2} h^{-1} for the first 5 days and 229 L m^{-2} h^{-1} for the remainder of the run. No fouling occurred and the TMP was below 0.5 psi (2.5 kPa) throughout the 14 days of operation.

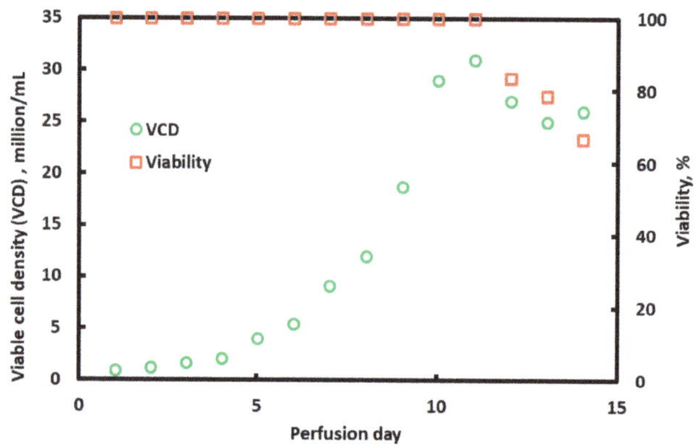

Figure 8. Variation in viable cell density and cell viability over a 14-day perfusion experiment. A BioOptimal MF SL microfilter was run in ATF mode as the cell retention device.

3.5. Integrated Hydrocyclone TFF

As shown in Figure 3c, an integrated hydrocyclone–TFF system was tested where the overflow from the hydrocyclone was used as the feed to the TFF module. Figure 9 gives the variation in the permeate flux with time. Figure 9a gives the results for the BioOptimal filter by itself and for the same module integrated with the hydrocyclone. Figure 9b gives the analogous results for the UJP module. As can be seen, the BioOptimal displayed a much higher permeate flux than the UJP module with and without an integrated hydrocyclone. Further, the rate of flux decline over a 2 h period was much greater for the UJP module. The cell viability was 97% at the start of the 2 h run and remained at approximately 97% at the end of the run for both modules. The much more stable flux for the BioOptimal module is similar to the results shown in Figure 5 and indicates the benefit of using a reverse asymmetric membrane in this application.

Figure 9a suggests that the flux decline for the BioOptimal was similar with and without an integrated hydrocyclone. The small difference in flux for the two runs was likely due to a small loss in the permeability of the module. The module was tested without the hydrocyclone; it was cleaned/regenerated at the end of the experiment and then used again for the integrated process. The permeability of the regenerated membrane was within 90% of the initial water flux through the clean module, but the small decrease in permeability likely caused the slightly lower flux for the integrated process. These results indicate that, for the BioOptimal module, an integrated hydrocyclone had little effect on membrane fouling under the operating conditions investigated, likely due to the effectiveness of the open pore structure in protecting the size-selective skin on the outer surface of the membrane from fouling by cells.

The results for the UJP module are shown in Figure 9b. Again, the initial flux was a little lower for the integrated process due to the reduction in the permeability of the regenerated module. However, the rate of flux decline for the integrated clarification process was significantly less than that for the UJP module alone. In this case, the ratio of the flux at $t = 40$ min to that at $t = 10$ min was 0.25 for the UJP module alone compared to 0.39 for the same module, but after primary clarification using the hydrocyclone. The net result is that the flux for the integrated hydrocyclone–TFF process was greater than that for the TFF alone for $t > 30$ min. These results suggest that, by using a hydrocyclone for initial clarification, the rate of flux decline due to membrane fouling can be reduced for conventional TFF microfiltration membranes in which the barrier layer faces the feed.

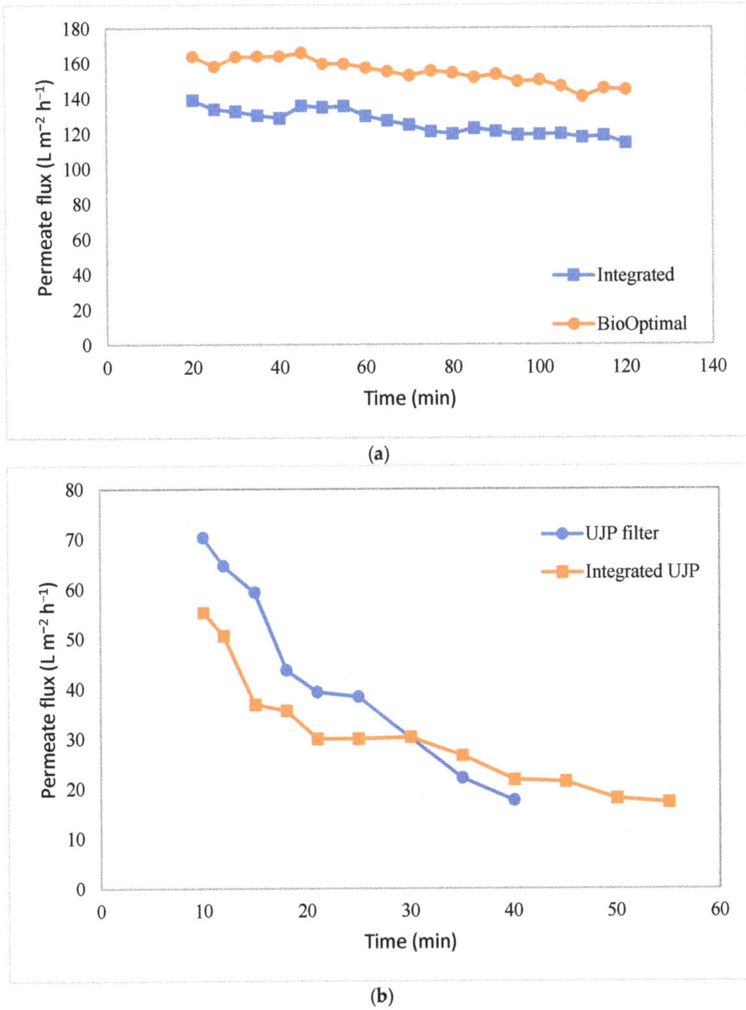

Figure 9. Variation in permeate flux with time: (**a**) BioOptimal and hydrocyclone–BioOptimal system and (**b**) UJP and hydrocylone–UJP system. Data for the integrated systems were obtained with the same module used for the TFF alone, but after cleaning/regeneration.

4. Discussion

TFF is a very attractive unit operation for clarification of cell culture media in perfusion bioreactors, as the permeate can typically be introduced directly to the first capture chromatography operation without any additional clarification. Furthermore, TFF can be easily adapted for both batch and continuous clarification. However, membrane fouling is a serious limitation that restricts the practical use of TFF. Figure 5 indicates that, for traditional TFF membranes, where the barrier layer faces the feed stream, rapid flux decline occurs due to absorption of foulants and deposition of rejected species directly on or within the barrier layer of the membrane. As shown in Figure 5 and in our earlier studies [7,8], the use of a reverse asymmetric membrane can be beneficial in this application, as the support structure helps stabilize a highly permeable layer of rejected particulate matter. Foulants are removed by this dynamic membrane, thus protecting the barrier layer that faces the permeate stream from membrane fouling. However, as seen in Figure 5, as the cell viability decreases, cell debris will cause significant fouling even of these reverse asymmetric membranes, leading to a more rapid flux decline and the need to replace the membrane module to enable long-term continuous operations. In a continuous process, the flux would be much lower, but must be maintained for up to 2 months.

Here, for the first time, we combined a hydrocyclone with TFF for clarification of a CHO cell culture. The hydrocyclone provides the primary clarification operation, significantly reducing the particulate concentration in the feed to the TFF module. For traditional TFF membranes where the barrier layer faces the feed, reducing the foulant load in the feed leads to more stable fluxes over a longer time (see Figure 9b). Note that these experiments were not run under sterile conditions, which limited the duration of the integrated clarification process. It is likely that, if the integrated hydrocyclone–TFF experiments were conducted under sterile conditions for several days, the more stable permeate flux for the integrated process would be more significant when compared to the TFF operation alone.

In contrast, at least over a 2 h period, it appears that the integrated process offers no measurable advantage when using the BioOptimal module for the second clarification step. A similar result was observed in parallel studies, where a BioOptimal module was used in ATF mode for 2 weeks. In this experiment, the permeate flux and the rate of medium addition to the bioreactor were matched: 152 L m^{-2} h^{-1} for the first 5 days of operation and then 229 L $m^{-2}h^{-1}$ for days 6 to 14 of operations. Both the traditional TFF membranes and the reverse asymmetric membrane showed no flux decline over a 2-week period. ATF has been proposed as another way to suppress membrane fouling. This method typically uses slightly larger-inner-diameter hollow fiber membranes, although any TFF membrane could potentially be used. ATF makes use of altered fluid mechanics and operation at a very low filtrate flux to suppress membrane fouling. A diaphragm pump is used that removes and returns the feed through the same port [21,22]. While feed flow reversal leads to suppression of fouling, a disadvantage of the process is that the permeate fluxes are low, resulting in very large membrane surface areas for large-scale commercial operation.

The results obtained here indicate that, for a reverse asymmetric membrane that is more resistant to flux decline, the benefits of an integrated process are likely to be greater at a low cell viability or for extremely long run times, e.g., 2 months. An integrated process may be well suited for continuous biomanufacturing operations, particularly when using conventional TFF membranes in which the size-selective layer faces the cell culture media. Further investigations under sterile conditions for much longer run times will be required in order to determine the effectiveness of integrating a hydrocyclone as the primary clarification step prior to TFF. Nevertheless, the membrane choice will be important (pore size, morphology, asymmetric structure, etc.) for the design and implementation of an optimized two-step clarification process incorporating both a hydrocyclone and a TFF membrane unit.

5. Conclusions

Developing methods to improve the performance of TFF for bioreactor clarification is highly desirable, as it can be adapted for both batch and continuous operations. Furthermore, the careful selection of the membrane pore size will enable the direct introduction of the permeate to the subsequent capture step in the downstream process. There have been many attempts to overcome membrane fouling, which is a major limitation that leads to product sieving losses, low permeate fluxes, and the need to periodically replace/regenerate the membrane. Here, the feasibility of integrating a hydrocyclone with TFF was investigated for the first time. Like TFF, a hydrocyclone can also be adapted for batch and continuous operations. The overflow from the hydrocyclone is depleted of large particulate matter, thus reducing the fouling potential of the feed in the TFF module. Care will be needed when integrating these two unit operations, as the overflow rate will determine the feed flow rate and wall shear rate for the TFF module. Furthermore, selecting an appropriate membrane morphology will be important. The results obtained here indicate that, for reverse asymmetric membranes that are much more resistant to fouling, integrating a hydrocyclone may be of minimal benefit, while significant improvements in the TFF performance may be possible when using more conventional asymmetric microfiltration membranes.

Author Contributions: Conceptualization, X.Q., J.P.S., S.R.W. and A.Z.; methodology, L.-K.C., A.H.M., X.Q., J.P.S., S.R.W. and A.Z.; investigation, L.-K.C. and A.H.M.; resources, X.Q., S.R.W. and A.Z.; data curation, A.H.M., S.R.W. and D.Z.; writing—original draft preparation, A.H.M.; writing—review and editing, A.H.M., S.R.W. and A.Z.; supervision, X.Q., S.R.W. and A.Z.; project administration, X.Q., S.R.W. and A.Z.; funding acquisition, X.Q., S.R.W. and A.Z. All authors have read and agreed to the published version of the manuscript.

Funding: Funding for this work was provided by the National Science Foundation Industry and University Cooperative Research Center for Membrane Applications Science and Technology (MAST) via grant numbers 1822101 and 1841474.

Institutional Review Board Statement: Not applicable.

Informed Consent Statement: Not applicable.

Data Availability Statement: The original contributions presented in this study are included in the article. Further inquiries can be directed to the corresponding authors.

Acknowledgments: The authors would like to acknowledge April Wheeler and Daniel Strauss, Asahi Kasei Bioprocess America, Inc., 1855 Elmdale Avenue, Glenview, IL 60026 USA, and Hironobu Shirataki, Asahi Kasei Medical, Chiyoda, Tokyo 100-0006, Japan, for their insights and assistance. The authors would also like to acknowledge Abdullah Loman, Biogen, 5000 Davis Dr, Research Triangle Park, NC 27709, USA, for his helpful insights and suggestions.

Conflicts of Interest: L.-K.C. was employed by Merck & Co., Inc.; D.Z. was employed by Benitec Biopharma Inc.; and J.P.S. was employed by Biogen. The remaining authors declare that this research was conducted in the absence of any commercial or financial relationships that could be construed as a potential conflict of interest.

References

1. Khanal, O.; Lenhoff, A.M. Developments and Opportunities in Continuous Biopharmaceutical Manufacturing. *MAbs* **2021**, *13*, e1903664. [CrossRef] [PubMed]
2. Zydney, A.L. New Developments in Membranes for Bioprocessing—A Review. *J. Memb. Sci.* **2021**, *620*, 118804. [CrossRef]
3. Drobnjakovic, M.; Hart, R.; Kulvatunyou, B.; Ivezic, N.; Srinivasan, V. Current Challenges and Recent Advances on the Path towards Continuous Biomanufacturing. *Biotechnol. Prog.* **2023**, *39*, e3378. [CrossRef]
4. Yang, O.; Qadan, M.; Ierapetritou, M. Economic Analysis of Batch and Continuous Biopharmaceutical Antibody Production: A Review. *J. Pharm. Innov.* **2020**, *15*, 182–200. [CrossRef] [PubMed]
5. Castilho, L.R.; Medronho, R.A. Cell Retention Devices for Suspended-Cell Perfusion Cultures. *Adv. Biochem. Eng. Biotechnol.* **2002**, *74*, 129–169. [CrossRef]
6. Van Reis, R.; Zydney, A. Membrane Separations in Biotechnology. *Curr. Opin. Biotechnol.* **2001**, *12*, 208–211. [CrossRef] [PubMed]
7. Zhang, D.; Patel, P.; Strauss, D.; Qian, X.; Wickramasinghe, S.R. Modeling Tangential Flow Filtration Using Reverse Asymmetric Membranes for Bioreactor Harvesting. *Biotechnol. Prog.* **2021**, *37*, e3084. [CrossRef] [PubMed]
8. Zhang, D.; Wickramasinghe, R.; Martin, R.E.; Patel, P.; Strauss, D.; Qian, X. Tangential Flow Filtration Using Reverse Asymmetric Membrane for CHO Cell Harvesting. *Biotechnol. Prog.* **2020**, *37*, e3115.
9. Walther, J.; McLarty, J.; Johnson, T. The Effects of Alternating Tangential Flow (ATF) Residence Time, Hydrodynamic Stress, and Filtration Flux on High-Density Perfusion Cell Culture. *Biotechnol. Bioeng.* **2019**, *116*, 320–332. [CrossRef]
10. Boodhoo, K.V.K.; Flickinger, M.C.; Woodley, J.M.; Emanuelsson, E.A.C. Bioprocess Intensification: A Route to Efficient and Sustainable Biocatalytic Transformations for the Future. *Chem. Eng. Process. Process Intensif.* **2022**, *172*, 108793. [CrossRef]

11. Müller, D.; Klein, L.; Lemke, J.; Schulze, M.; Kruse, T.; Saballus, M.; Matuszczyk, J.; Kampmann, M.; Zijlstra, G. Process Intensification in the Biopharma Industry: Improving Efficiency of Protein Manufacturing Processes from Development to Production Scale Using Synergistic Approaches. *Chem. Eng. Process. Process Intensif.* **2022**, *171*, 108727. [CrossRef]
12. Tian, J.; Ni, L.; Song, T.; Olson, J.; Zhao, J. An Overview of Operating Parameters and Conditions in Hydrocyclones for Enhanced Separations. *Sep. Purif. Technol.* **2018**, *206*, 268–285. [CrossRef]
13. Ni, L.; Tian, J.; Song, T.; Jong, Y.; Zhao, J. Optimizing Geometric Parameters in Hydrocyclones for Enhanced Separations: A Review and Perspective. *Sep. Purif. Rev.* **2018**, *48*, 30–51. [CrossRef]
14. Ji, L.; Paul, P.; Shanbhag, B.K.; Dixon, I.; Kuang, S.; He, L. Emerging Application of Hydrocyclone in Biotechnology and Food Processing. *Sep. Purif. Technol.* **2023**, *309*, 122992. [CrossRef]
15. Elsayed, E.A.; Medronho, R.A.; Wagner, R.; Deckwer, W.D. Use of Hydrocyclones for Mammalian Cell Retention: Separation Efficiency and Cell Viability (Part 1). *Eng. Life Sci.* **2006**, *6*, 347–354. [CrossRef]
16. Bettinardi, I.W.; Castan, A.; Medronho, R.A.; Castilho, L.R. Hydrocyclones as Cell Retention Device for CHO Perfusion Processes in Single-Use Bioreactors. *Biotechnol. Bioeng.* **2020**, *117*, 1915–1928. [CrossRef] [PubMed]
17. Syed, M.S.; Marquis, C.; Taylor, R.; Warkiani, M.E. A Two-Step Microengineered System for High-Density Cell Retention from Bioreactors. *Sep. Purif. Technol.* **2021**, *254*, 117610. [CrossRef]
18. He, L.; Ji, L.; He, Y.; Liu, Y.; Chen, S.; Chu, K.; Kuang, S. Experimental and Numerical Analysis of Chinese Hamster Ovary Cell Viability Loss in Mini-Hydrocyclones. *Sep. Purif. Technol.* **2022**, *295*, 121203. [CrossRef]
19. Castilho, L.R.; Medronho, R.A. A Simple Procedure for Design and Performance Prediction of Bradley and Rietema Hydrocyclones. *Miner Eng.* **2000**, *13*, 183–191. [CrossRef]
20. Bradley, D. *The Hydrocyclone: International Series of Monographs in Chemical Engineering*; Pergamon Press Ltd.: London, UK, 1965.
21. Clincke, M.F.; Mölleryd, C.; Samani, P.K.; Lindskog, E.; Fäldt, E.; Walsh, K.; Chotteau, V. Very High Density of Chinese Hamster Ovary Cells in Perfusion by Alternating Tangential Flow or Tangential Flow Filtration in WAVE Bioreactor™—Part II: Applications for Antibody Production and Cryopreservation. *Biotechnol. Prog.* **2013**, *29*, 768–777. [CrossRef]
22. Kelly, W.; Scully, J.; Zhang, D.; Feng, G.; Lavengood, M.; Condon, J.; Knighton, J.; Bhatia, R. Understanding and Modeling Alternating Tangential Flow Filtration for Perfusion Cell Culture. *Biotechnol. Prog.* **2014**, *30*, 1291–1300. [CrossRef] [PubMed]

membranes

MDPI

Article

Enhancing Virus Filter Performance Through Pretreatment by Membrane Adsorbers

Solomon Isu [1,†], Shu-Ting Chen [1], Raheleh Daneshpour [1,‡], Hironobu Shirataki [2], Daniel Strauss [3], Andrew L. Zydney [4], Xianghong Qian [5,*] and Sumith Ranil Wickramasinghe [1,*]

[1] Ralph E. Martin Department of Chemical Engineering, University of Arkansas, Fayetteville, AR 72701, USA; solomon.isu@milliporesigma.com (S.I.); sc068@uark.edu (S.-T.C.); daneshpour@psu.edu (R.D.)
[2] Scientific Affairs Group, Bioprocess Division, Asahi Kasei Medical, Chiyoda, Tokyo 100-0006, Japan; shirataki.hb@om.asahi-kasei.co.jp
[3] Research and Development, Asahi Kasei Bioprocess America, Glenview, IL 60026, USA; daniel.strauss@ak-bio.com
[4] Department of Chemical Engineering, The Pennsylvania State University, University Park, PA 16802, USA; zydney@engr.psu.edu
[5] Department of Biomedical Engineering, University of Arkansas, Fayetteville, AR 72701, USA
* Correspondence: xqian@uark.edu (X.Q.); ranil.wickramasinghe@uark.edu (S.R.W.)
† Current address: Technical and Scientific Solutions, EMD Millipore Corporation, Burlington, MA 01803, USA.
‡ Current address: Department of Chemical Engineering, The Pennsylvania State University, University Park, PA 16802, USA.

Abstract: Virus filtration is used to ensure the high level of virus clearance required in the manufacture of biopharmaceutical products such as monoclonal antibodies. Flux decline during virus filtration can occur due to the formation of reversible aggregates consisting of self-assembled monomeric monoclonal antibody molecules, particularly at high antibody concentrations. While size exclusion chromatography is generally unable to detect these reversible aggregates, dynamic light scattering may be used to determine their presence. Flux decline during virus filtration may be minimized by pretreating the feed using a membrane adsorber in order to disrupt the reversible aggregates that are present. The formation of reversible aggregates is highly dependent on the monoclonal antibody and the feed conditions. For the pH values investigated here, pretreatment of the feed using a hydrophobic interaction membrane adsorber was the most effective in minimizing flux decline during virus filtration. Ion exchange membranes may also be effective if the monoclonal antibody and membrane are oppositely charged. Consequently, the effectiveness of ion exchange membrane adsorbers is much more dependent on solution pH when compared to hydrophobic interaction membrane adsorbers. Size based prefiltration was found to be ineffective at disrupting these reversible aggregates. These results can help guide the development of more effective virus filtration processes for monoclonal antibody production.

Keywords: aggregation; flux decline; fouling; membrane adsorber; monoclonal antibody; pH; prefiltration; reversible aggregate; virus filtration

1. Introduction

Virus filtration is routinely used in the manufacture of protein-based therapeutics such as monoclonal antibodies (mAbs). Validation of adequate virus clearance is critical in the manufacture of biopharmaceutical products. Virus filtration is a robust unit operation that makes use of membrane-based size exclusion to reject potentially contaminating virus particles. Virus filters are often characterized based on their rejection of model virus particles, typically minute virus of mice (MVM). The MVM capsid is about 25 nm in diameter and 2 nm thick and is arranged with a T = 1 icosahedral symmetry [1]. Virus filters are operated in normal flow mode under constant pressure. They are designed to provide at least 10,000-fold decrease in virus concentration in the permeate relative to the feed (4 log removal of virus, LRV). Additionally for the process to be economically viable, more than 95% product recovery is required. Design limits on product throughput (minimum product recovered per membrane surface area) and permeate flux put further constraints on the design of virus filters [2].

The size of antibodies in solution is highly variable and depends on buffer conditions. Roughly, biophysical analysis indicates they are around 12 nm long, with 3 rod-shaped arms each about 3.5 nm in diameter [3]. As membrane based size exclusion is usually most effective when the difference in size of the species to be separated is around an order of magnitude, the design criteria for virus filtration membranes is very demanding [4]. Since virus filtration occurs towards the end of the purification train, fouling of virus filters is most commonly due to product related contaminants such as product aggregates, dimers, trimers, and misfolded product molecules instead of host cell proteins [5].

Several recent studies have investigated an integrated 'pretreatment' step using either column chromatography or a virus prefilter to minimize fouling of the virus filtration membrane and improve the flux and throughput. These studies have indicated that the presence of product aggregates that are less than 100 nm in size are commonly responsible for the decrease in flux during virus filtration. While the feed stream is usually passed through a 0.2 or 0.1 μm sterilizing grade filter prior to introduction to the virus filter, smaller aggregates will not be removed and can then block the pores of the virus filter [6]. Further sized based prefiltration will not remove foulants that are similar in size to the product species. These foulants can potentially be removed by surface interactions. Brown et al. [7] used ion exchange membrane adsorbers to remove product related aggregates 8–13 nm in size. Rayfield et al. [8] indicated the importance of feed properties such as pH, buffer type, salt type, and concentration, all of which affect the biophysical properties of the mAb. This, in turn, can lead to greater aggregation and product related contaminants. Stanevich et al. [9] investigated the use of polyamide 6-6 (nylon 6-6) membranes as adsorptive prefilters, which function through hydrophobic interactions. Bolton et al. [10] used a prefilter that contained diatomaceous earth to bind aggregates through hydrophobic interactions. Others, such as Shirataki et al. [11], have used mixed mode resins as a pretreatment step prior to virus filtration.

Johnson et al. [12] provide a recent summary of current industrial practices in virus filtration, including the use of prefiltration. In another recent study, Isu et al. [13] describe the use of bioanalytical methods to identify aggregation-prone and less filterable proteoforms during virus filtration, which could guide the selection of an appropriate pretreatment strategy. Several investigators have noted that severe fouling is often observed even after prefiltration [5]. They note that this is due to adsorption of the mAb onto the virus filtration membrane surface. Furthermore, some investigators have noted severe membrane fouling even when the amount of high molecular weight species is less than 1% in the feed [8]. This is likely due to reversible self-association of the monomeric mAb [14].

Bieberbach et al. [15] indicate that these observations may be explained by considering the formation of irreversible and reversible aggregates [16–18]. Irreversible aggregates can be soluble or insoluble. They are held together by strong intermolecular forces such as covalent bonds. They can generally be detected by size exclusion chromatography. Reversible protein aggregation or self-assembly leads to the formation of soluble oligomers. They are held together by weak intermolecular interactions and are more likely to form at high concentrations that would exist in the narrow pores of a virus filter. These reversible aggregates may revert to protein monomers upon dilution [19]. Protein self-assembly is a concern in the manufacture and storage of biopharmaceutical products. Dynamic light scattering may be used to qualitatively determine the degree of protein self-assembly in solution [20].

The observations described above are further complicated by the fact that virus filters from different manufacturers are made of different polymeric materials [8]. Proprietary surface treatments are used to make the membrane surface more biocompatible and to suppress fouling. However, the degree of adsorption of a specific mAb, including both self-assembled mAb species as well as irreversible aggregates, will depend on the mAb, the specific surface properties of the membrane, and the buffer conditions.

In this study, we have investigated the performance of the Planova BioEx virus removal filter (Asahi Kasei Bioprocess, Glenview, IL, USA) using amAb under different pH conditions. Size based membranes, as well as ion exchange and hydrophobic interaction membrane adsorbers, have been used as a prefiltration/pretreatment step. The impact of these pretreatment steps on the permeate flux through the virus filter has been determined. Dynamic light scattering has been used to determine the level of self-assembly of the mAb under different pH conditions. Three pH conditions were considered: above (pH 8.6), around (pH 7.5), and below (pH 5.0) the isoelectric point of the mAb. Due to the stability of the mAb, pH values much above the isoelectric point of the mAb could not be investigated. While the pH values investigated are of relevance industrially, they also enable investigation of the behavior of the mAb at pH values below, near, and above its isoelectric point. The results provide guidelines for the selection of a prefilter/membrane adsorber to be used prior to virus filtration.

2. Materials and Methods

2.1. Reagents and Filters

The reagents used for preparing the various buffers were sodium acetate trihydrate (molecular biology grade, >99% purity), sodium chloride, sodium phosphate monobasic monohydrate (ACS reagent, >98% purity), sodium phosphate dibasic (reagent plus, >99% purity), and glacial acetic acid from MilliporeSigma (Billerica, MA, USA). Ammonium sulfate was obtained from VWR Life Science (Radnor, PA, USA). Ultrapure water with a resistivity of 18.2 MΩ was used for buffer formulation.

The mAb used in these studies had a molecular weight of 148 kDa and an isoelectric point of 8.0. Four different feed buffer solutions were prepared consisting of 20 mM sodium acetate at pH values of 5.0, 7.5, and 8.6. The pH was adjusted using glacial acetic acid. At pH 5.0 and 8.6, 200 mM NaCl was added. At pH 7.5 the buffer was prepared with and without 200 mM NaCl. The mAb concentration in all solutions was 5 g L^{-1}.

The prefilters and membrane adsorbers used in this study are summarized in Table 1. In this work, the prefilter or membrane adsorber was not installed in-line with the virus filter. Instead, the pretreated product solution was introduced to the virus filter within 5 min of prefiltration. The virus filter used was the Planova BioEx with membrane surface area 0.0003 m^2 (Asahi Kasei Bioprocess, Glenview IL, USA). All feed streams were prefiltered through a 0.2 μm bottle top filter. Results for the 0.2 μm bottle top filter did not include any further pretreatment. Results for all other prefilters and membrane adsorbers were generated for feed streams after passage through the 0.2 μm bottle top filter. The feed from the 0.2 μm bottle top filter was used within 5 min either as the feed to the virus filter or the feed to the subsequent pretreatment step.

Table 1. Details of prefilters and membrane adsorbers used in this study.

Prefilter	Membrane Material	Source	Feed Streams Tested	Mechanism of Action	Comment
0.2 µm bottle top filter	Polyethersulfone	ThermoFisher Scientific (Waltham, MA, USA)	Acetate buffer with 200 mM NaCl at pH 5.0, 7.5 and 8.6. Acetate buffer with no added NaCl at pH 7.5.	Size exclusion	Used for all feed streams. Tested by itself for feed stream at pH 5.0 and 8.6 with 200 mM NaCl; pH 7.5 without 200 mM NaCl.
0.1 µm bottle top filter	Polyethersulfone	ThermoFisher Scientific (Waltham, MA, USA)	Acetate buffer with 200 mM NaCl at pH 5.0.	Size exclusion	Filtered through 0.2 µm bottle top filter, then through 0.1 µm bottle top filter.
Planova 75 N (75 N)	Regenerated cellulose hollow fibers, 0.001 m^2 surface area, nominal pore size 75 nm	Asahi Kasei Bioprocess (Glenview, IL, USA)	Acetate buffer with 200 mM NaCl at pH 5.0.	Size exclusion	Filtered through 0.2 µm bottle top filter, then through Planova 75 N filter.
Sartobind Q (nano) (IEX-Q)	Cellulosic membrane, quaternary ammonium ligands, 3 mL bed volume, 8 mm bed height, pore size > 3 µm.	Sartorius AG, (Göttingen, Germany)	Acetate buffer with 200 mM NaCl at pH 5.0 and 8.6. Acetate buffer with no added NaCl at pH 7.5.	Anion exchange	Filtered through 0.2 µm bottle top filter, then through Sartobind Q adsorber.
Sartobind S (nano) (IEX-S)	Cellulosic membrane, sulfonic ligands, 3 mL bed volume, 8 mm bed height, pore size > 3 µm.	Sartorius AG, (Göttingen, Germany)	Acetate buffer with 200 mM NaCl at pH 5.0 and pH 8.6. Acetate buffer with no added NaCl at pH 7.5.	Cation exchange	Filtered through 0.2 µm bottle top filter, then through Sartobind S adsorber.
Sartobind Phenyl (HIC)	Cellulosic membrane, phenyl ligands, 3 mL bed volume, 8 mm bed height, pore size > 3 µm.	Sartorius AG, (Göttingen, Germany)	Acetate buffer with 200 mM NaCl at pH 5.0, 7.5, 8.6.	Hydrophobic interaction	Filtered through 0.2 µm bottle top filter, then through Sartobind Phenyl adsorber.

Membrane adsorbers were installed on a fast protein liquid chromatography (FPLC) system (Cytiva, Marlborough, MA, USA). For hydrophobic interaction chromatography (HIC) membrane adsorber pretreatment (at all pH values 5.0, 7.5, 8.6), the buffer included 200 mM NaCl. For ion exchange membrane adsorbers, 200 mM NaCl was present in the buffer at pH 5.0 and 8.6 only. For the sized based prefilters (0.1 µm bottle top filter and 75 N), no NaCl was added to the feed.

2.2. Analytical Methods

Buffer pH and conductivity were measured using an Orion Star™ pH/conductivity benchtop multiparameter meter from ThermoFisher Scientific (Waltham, MA, USA). The mAb concentration was measured by UV absorbance at 280 nm using a Genesys10 UV Scanning System (Waltham, MA, USA) with VWR Quartz Spectrophotometer Cell (path length 1 cm). Dynamic light scattering (DLS) was performed using a Delsa™ Nano particle size analyzer from Beckman Coulter (Brea, CA, USA) to determine the hydrodynamic diameter of the mAb. Results represent the average of three readings.

Size exclusion chromatography (SEC) was conducted to determine the aggregate fraction in the mAb solution at pH 5.0 after prefiltration through the 0.2 μm bottle top filter, after filtration through the BioEx, and in the buffer flush performed after mAb filtration. A TSKgel G3000SWXL column with 7.8 mm ID, 30 cm length, and particle size of 5 μm (Tosoh Bioscience, Grove City, OH, USA) was used. The column was installed on a high-performance liquid chromatography (HPLC) instrument, Agilent 1260 Infinity Quaternary LC, (Agilent Technologies, Santa Clara, CA, USA). The mobile phase was 20 mM sodium phosphate buffer, pH 7.0, with 300 mM ammonium sulfate. The SEC column was equilibrated with the mobile phase at 0.9 mL min^{-1} for one hour before sample introduction. All samples were filtered through a 0.2-μm syringe filter and loaded into 1 mL sample vials. The HPLC run cycle was twenty minutes at a flow rate of 0.9 mL min^{-1}. Subsequently, 10 μL samples were injected into the column and analyzed.

2.3. Virus Filtration

During virus filtration, the weight of the permeate was recorded. Visual leak integrity tests (as described by the manufacturer) were performed on the BioEX virus filter at 100 kPa for 20 s before flushing air out of the system. Deionized water filtered with a 0.2 μm bottle top filter was added into the Planova™ Pressure Reservoir (Asahi Kasei Bioprocess, Glenview, IL, USA). The virus filter was flushed with 40 L m^{-2} of DI water and then with 40 L m^{-2} of the feed buffer. The mAb loading volume was about 65 mL (250 L m^{-2}) at a feed pressure of 310 kPa. For feed streams prefiltered with the 0.2 μm bottle top filter only, a buffer flush was included.

3. Results

In this study, 200 mM NaCl was not added to all the feed streams at pH 7.5, which was close to the isoelectric point of the protein. This was performed to minimize aggregate formation [21,22]. When the pH of the solution is close to the isoelectric point of the mAb, a high ionic strength buffer can cause 'salting-out', which could lead to increased aggregate formation. However, 200 mM NaCl was added to all feed streams that were pretreated with the HIC membrane adsorber as adsorption of the impurities onto the hydrophobic ligands is enhanced at high salt concentrations.

Dynamic light scattering was used to determine the diameter of the mAb for all feed streams after prefiltration. Table 2 summarizes the results. As can be seen, there is considerable variability in the diameter of the mAb. Without additional pretreatment beyond the 0.2 μm prefilter, the average diameter was largest for the mAb at pH 5.0. The hydrodynamic diameter was lowest close to the isoelectric point of the mAb, even though these conditions would be expected to minimize electrostatic repulsive interactions. Interestingly, the diameter was larger for the pH 5.0 feed prefiltered through the Planova 75 N than that filtered through the 0.1 and 0.2 μm prefilters, even though the 75 N has a significantly smaller pore size.

Table 2. Average mAb diameter measured by dynamic light scattering for the various feed streams after pretreatment.

Pre-Filter	Average Diameter in Permeate (nm)	Buffer Condition
IEX-S	12.1	pH 5.0
HIC	11.1	
IEX-Q	15.5	
0.1 μm	17.4	
0.2 μm	21.4	
75 N	22.1	
IEX-S	12.4	pH 7.5
HIC	12.5	
IEX-Q	13.6	
0.2 μm	12.1	
IEX-S	15.4	pH 8.6
HIC	14.2	
IEX-Q	14.4	
0.2 μm	15.8	

Figure 1 gives the virus filter flux as a function of throughput for the three pH values tested for the feed stream pretreated only using a 0.2 μm bottle top filter. Though the aim was to load the virus filter to 250 L m^{-2}, due to rapid flux decline this was not achieved. The filtration was stopped when the flux dropped below 40 L m^{-2} h^{-1}, at which point the buffer flush using 20 mL of the feed buffer was commenced. As can be seen, the flux increases somewhat during the buffer flush. The flux decline is most rapid at pH 5.0, as is the flux recovery during the buffer flush. The flux decline appears to be least at pH 7.5, which is close to the isoelectric point of the mAb, with the flux remaining >40 L m^{-2} h^{-1} out to a throughput of 230 L m^{-2}.

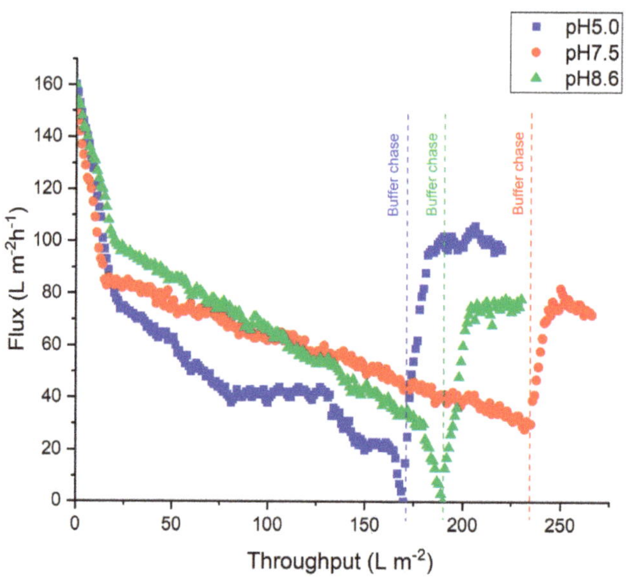

Figure 1. Variation of virus filter flux versus throughput for feed streams at different pH after pretreatment with a 0.2 μm bottle top filter. When flux values dropped to below 40 L m^{-2} h^{-1}, a buffer flush was initiated and the flux increased.

Since the feed stream at pH 5.0 displayed the most rapid flux decline, samples of the feed and permeate after the virus filtration and after the buffer flush were analyzed by SEC. The results are shown in Table 3. It can be seen that the percentage of dimer is very low in all samples. No higher molecular weight oligomers were detected.

Table 3. SEC data for the feed and permeate after virus filtration and buffer flush. The feed pH was 5.0.

	Feed	Permeate: Virus Filtration	Permeate: Buffer Flush
Oligomer	0%	0%	0.5%
Dimer	5.2%	0%	1.4%
Monomer	94.8%	100%	98.1%

Figure 2 gives the permeate flux through the Planova BioEx virus filter as a function of throughput at pH 5.0 for feed streams pretreated using the various prefilters and membrane adsorbers listed in Table 1. All feed streams were passed through the 0.2 μm bottle top filter before further pretreatment. Results for pretreatment with the 0.2 μm bottle top filter only are given in Figure 1. As can be seen for all the different pretreatment methods, the flux through the virus filter decreases with time. The hydrophobic interaction membrane adsorbers performed the best as the decrease in flux is least for feed streams pretreated with this membrane adsorber, with the flux remaining above 95 L m^{-2} h^{-1} out to a throughput of 250 L m^{-2}. The next best performance was obtained after pretreatment with the cation exchange adsorber (IEX-S).

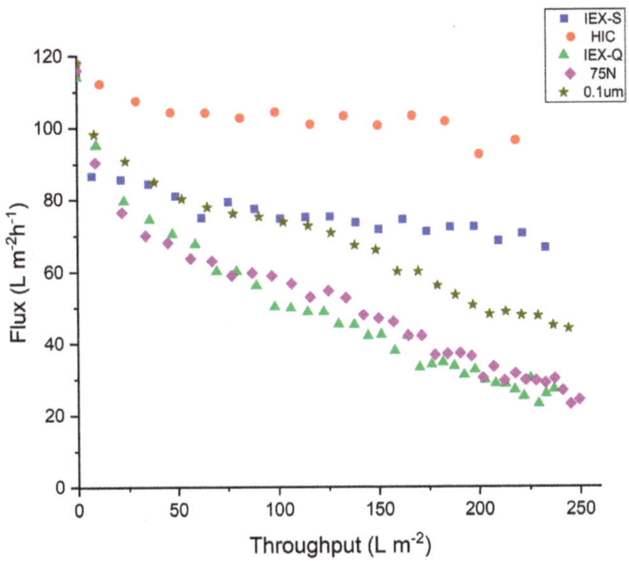

Figure 2. Variation of permeate flux through the BioEx virus filter versus throughput for feed streams pretreated using the various prefilters and membrane adsorbers listed in Table 1. The feed pH was 5.0.

Figure 3 shows analogous results to Figure 2 but for a feed pH of 7.5. While pretreatment using the IEX-S gave the highest flux, all three membrane adsorbers provided a stable flux over 250 L m^{-2} throughput. Note that the permeate flux through the feed pretreated with only the 0.2 μm bottle top filter (without any membrane adsorber) declines to <40 L m^{-2} h^{-1} after 250 L m^{-2} throughput under otherwise identical conditions.

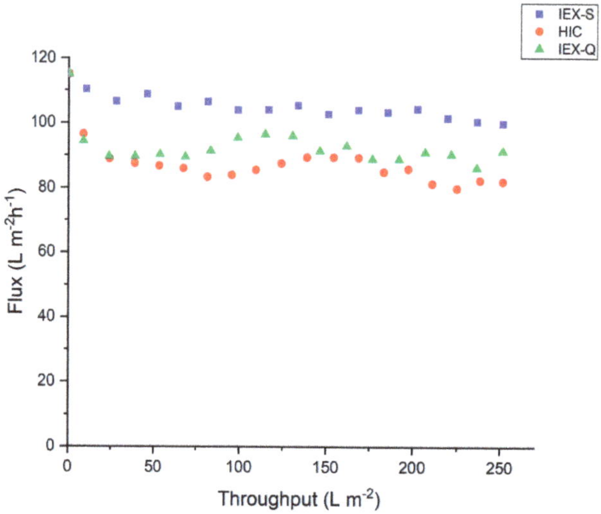

Figure 3. Variation of permeate flux through the BioEx virus filter versus throughput for feed streams pretreated using the various membrane adsorbers listed in Table 1. The feed pH was 7.5.

Finally, results for permeate flux through the BioEx filter for feed streams at pH 8.6 prefiltered using the three membrane adsorbers are given in Figure 4. As can be seen, a significant flux decay was observed throughout the run for all three conditions, with the least flux decline obtained using the HIC membrane adsorber. The results with the IEX-S adsorber were similar to those obtained in Figure 2 for the bottle top filter alone, suggesting that the negatively-charged adsorber is unable to remove any significant aggregates when the pH of the solution is above the pI of the antibody, i.e., under conditions where the antibody is negatively-charged.

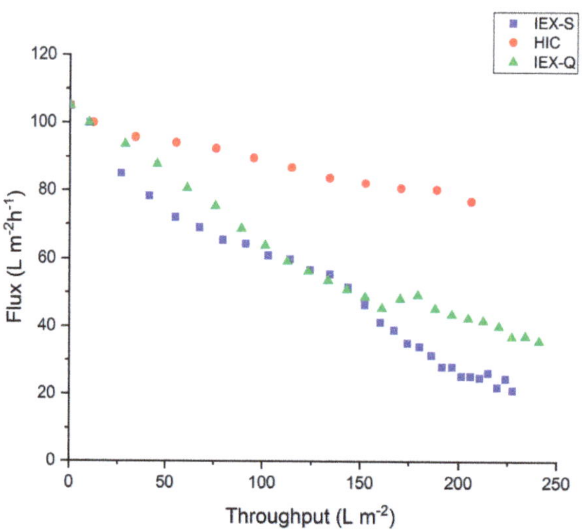

Figure 4. Variation of permeate flux through the BioEx virus filter versus throughput for feed streams pretreated using the various membrane adsorbers listed in Table 1. The feed pH was 8.6.

4. Discussion

The results in Figure 1 indicate that at pH values below, around, and above the isoelectric point of the mAb, rapid flux decline is observed during virus filtration when the feed stream is only prefiltered through a 0.2 μm filter. The SEC data (Table 3) for the feed stream at pH 5.0, which gave the most rapid flux decline during virus filtration, indicate the absence of any higher molecular weight oligomers while the percentage dimer is only around 5%. On the other hand, the DLS data (Table 2) indicate that the hydrodynamic diameter is the largest for the feed stream at pH 5.0 with an average size of 21.4 nm, which is essentially the same as the mean pore diameter of the BioEx virus removal filter. These data suggest that fouling is due to reversible aggregate formation, with these aggregates detected by DLS but not by SEC due to dilution of the protein feed by the buffer used in the SEC analysis. Further evidence of this is provided by the fact that there is rapid flux recovery during the buffer flush under all conditions, something that would not be expected if the fouling were due to irreversible aggregates that block the pores. Note that Billups et al. [2] also observed rapid flux recovery during the buffer flush due to dilution of the mAb and the dissociation of these reversible aggregates. The flux decline is intermediate at pH 8.6 and least at pH 7.5. The degree of flux decline is correlated to the measured hydrodynamic diameter (Table 2), with the smallest hydrodynamic diameter of 11.3 nm observed at pH 7.5.

Figure 2 indicates that rapid flux decline is observed at pH 5.0 for the feed streams that were prefiltered with just one of the size based prefilters. Interestingly, the flux decline is less pronounced for the feed processed through the 0.1 μm pore size prefilter than the 75 N even though the 75 N has a smaller nominal pore size. This difference may be associated with the different chemistry of these prefilters (Table 1); the 0.1 μm prefilter is a polyethersulfone membrane while the 75 N is regenerated cellulose. The more hydrophobic polyethersulfone membrane may provide better removal of foulants by adsorptive interactions, consistent with the strong performance of the HIC membrane adsorber in arresting the flux decline during virus filtration. Thus, it appears that membrane hydrophobicity plays an important role in disrupting or removing the reversible aggregates that form at pH 5.0.

The IEX-Q and IEX-S membranes are cellulosic with quaternary amine and sulfonic acid ligands, respectively. At pH 5.0, the mAb will be positively charged. Thus, there will be a net repulsion from the quaternary amine ligands on the IEX-Q membrane while there will be a strong attractive interaction with the sulfonic acid ligands on the IEX-S membrane. This attractive interaction provides more effective disruption/removal of the reversible aggregates, leading to the greater effectiveness of the IEX-S membrane adsorber at arresting the flux decline during virus filtration.

The hydrodynamic diameter after pretreatment provides a qualitative indication of the likelihood of rapid flux decline during virus filtration. The feed stream after pretreatment with the HIC membrane adsorber had the smallest hydrodynamic diameter and, in turn, the highest and most stable flux during virus filtration.

The variation of virus filter flux at pH 7.5, which is close to the isoelectric point of the mAb, is very different than that observed at pH 5.0. Based on the results in Figures 2 and 3, it appeared that sized based prefiltration is not effective at arresting the decline in flux. Consequently, only the membrane adsorbers were investigated. All three adsorbers gave similar results, with relatively stable flux, consistent with the similar hydrodynamic diameter for these three cases (Table 2). Note that both the hydrodynamic diameter and fouling behavior were similar to that observed for the feed stream at pH 5.0, which was pretreated with the HIC membrane adsorber.

Protein aggregation (reversible and irreversible) is complex and highly dependent on the specific mAb. Generally, aggregation implies the formation of non-native oligomeric species by interactions between 'modified' monomers [23]. Reversible aggregation, however, refers to the noncovalent interactions between native monomers. Interactions between mAb monomers can be induced because of the charge distribution, charge heterogeneity, and surface hydrophobicity [24]. Esfandiary et al. [25] indicated that reversible aggregation or self-association consists of both enthalpic and entropic contributions. Electrostatic interactions, however, are more enthalpic in nature while hydrophobic interactions are more entropic in nature [26]. The electrostatic effect is generally more dominant. This would explain the much smaller hydrodynamic diameter when the solution pH is close to the mAb isoelectric point.

When the pH of the solution is close to the isoelectric point of the mAb, the net charge of the mAb will be close to zero. Thus, hydrophobic patches, as well as positive and negative charge patches, will be able to interact with the HIC, IEX-S, and IEX-Q membranes. It is also important to note that at pH 7.5, no NaCl was added to the feed nor to the IEX-S and IEX-Q membranes. The charge screening effect of higher ionic strength buffers will be absent, thus increasing possible interactions between the IEX ligands and the mAb. It appears that interactions between the membrane and mAb are important to suppress, at least temporarily, the reversible mAb aggregation. In the case of the 0.2 μm prefilter, the hydrophobic interactions will be much weaker as the feed buffer did not contain 200 mM salt, perhaps explaining the observed decline in permeate flux through the BioEx virus filter.

The observed decreases in permeate flux at pH 8.6, shown in Figure 4, are in general agreement with the above discussion. The hydrodynamic diameter of the mAb after pretreatment with the three membrane adsorbers is smallest for the HIC and IEX-Q membranes. It is largest for prefiltration with the 0.2 μm filter, which also shows the greatest flux decline during virus filtration (Figure 1). Although pretreatment with the HIC membrane adsorber leads to the highest permeate flux, a noticeable flux decline is observed compared to feed streams at pH 5.0 and 7.5. Furthermore, the hydrodynamic radius is larger at pH 8.6 than at pH 5.0 or 7.5. Reversible and irreversible mAb aggregation is highly dependent on solution conditions. Thus, it is not surprising that the hydrodynamic diameter after HIC pretreatment is larger at pH 8.6 compared to the lower pH values. At pH values greater than the isoelectric point of the mAb, the mAb will be negatively charged. Thus, it will interact more strongly with the quaternary ammonium ligands of the IEX-Q membrane adsorber. Table 2 indicates that the hydrodynamic diameter of the mAb is smaller after pretreatment with the IEX-Q membrane adsorber compared to the IEX-S membrane adsorber, although the flux decline is only slightly improved for the feed stream pretreated with the IEX-Q membrane adsorber compared to the IEX-S membrane adsorber.

Our results indicate that determining the hydrodynamic diameter of the mAb by DLS provides a qualitative indication of the likely flux decline. Figure 5 gives the variation of virus filter flux at 150 L m^{-2} with mAb hydrodynamic diameter for the various experiments conducted here. As can be seen, though, there is some scatter; a lower hydrodynamic diameter is a good indication of the absence of reversible aggregates and higher virus filter flux.

Taken together, our results indicate that pretreatment with the HIC membrane adsorber was generally the most effective in arresting the flux decline observed during virus filtration. Billups et al. [2] also noted that hydrophobic interactions between the mAb and a membrane placed directly on top and upstream of the virus filtration membrane was successful in disrupting reversible aggregates. Our results indicate that electrostatic interactions between the mAb and membrane can also lead to dissociation of reversible aggregates and an improvement in virus filter flux. However, the effectiveness of the ion exchange membrane adsorbers is much more pH sensitive, with significant reductions in flux decline only observed when the mAb and ion exchange ligands are oppositely charged.

A similar result was observed by Shirataki et al. [11] using mixed mode resins. Since the formation of reversible aggregates is highly dependent on solution conditions, the pH and ionic strength could potentially be adjusted to minimize the formation of reversible aggregates.

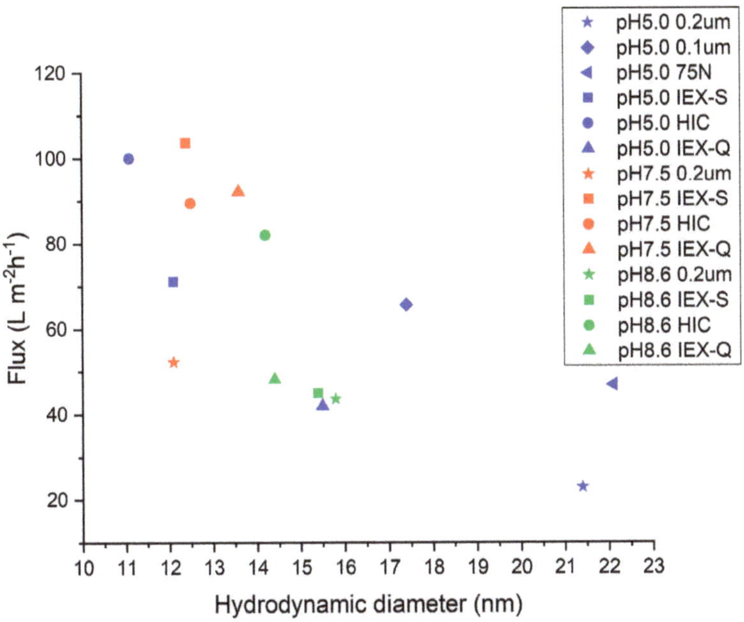

Figure 5. Variation of permeate flux through the BioEx virus filter versus hydrodynamic diameter using the various pretreatments listed in Table 2. Results are given for virus filter flux at 150 L m^{-2} throughput.

In addition, the formation of reversible aggregates is likely to be highly time dependent. Once disrupted, the reversible aggregates can reform with time. Consequently, using an in-line pretreatment step is likely the most effective approach to reduce fouling. It is possible that an in-line pretreatment step would be more effective than the separate pretreatment and virus filtration steps conducted here [7]. Further, while DLS measurements provide an indication of the degree of self-association of the mAb in solution, the environment in the constricted pores of a virus filter is very different and may well promote the formation of reversible aggregates due to the increase in local mAb concentration.

This investigation was conducted using only one mAb and at one mAb concentration, 5 g L^{-1}. While this value is representative of mAb concentrations used industrially, the formation of reversible aggregates is highly concentration dependent. In fact, as shown here, during the buffer chase due to dilution of the mAb, the reversible aggregates present are disrupted, leading to higher permeate fluxes during virus filtration. Future work will focus on the effect of mAb concentration on the disruption of the reversible aggregate and the effectiveness of pretreatment.

For ion exchange membrane adsorbers, charged patches on the aggregate surface interact with the oppositely charged ligands on the membrane surface. For hydrophobic interaction membrane adsorbers, hydrophobic patches on the aggregate surface interact with hydrophobic ligands present on the membrane surface. It is likely that if the ligands on the membrane adsorber interact with the reversible aggregates, it will lead to disruption of the reversible aggregates rather than adsorption given the weak interactions between mAb molecules. Further investigation is needed to determine the extent of reversible aggregate disruption versus adsorption.

It should be noted that besides the formation of reversible aggregates, flux decline during virus filtration after the feed has been prefiltered through a 0.2 or 0.1 μm filter can occur due to other product-related contaminants. Insoluble dimers formed from non-native mAb species can easily pass through a 0.2 or 0.1 μm filter. These dimers should be detectable by SEC. Since denatured mAb monomers are more hydrophobic than the native mAb monomers, the use of a HIC membrane adsorber may be effective. Jones et al. [27] indicate that certain 'high-risk' host cell proteins can also associate with the mAb product and thus pass through the purification train and copurify with the mAb. While these species can easily pass through the pores of a 0.2 or 0.1 μm prefilter, they can lead to fouling of the virus filter. These authors indicate various analytical methods such as mass spectrometry and HCP ELISA assays may be useful to detect the presence of these high-risk HCPs.

5. Conclusions

As virus filtration occurs towards the end of the purification train, fouling of virus filters is usually due to product related contaminants, including the presence of reversible aggregates. These aggregates can easily pass through a 0.2 or 0.1 μm prefilter, making these size-based prefilters largely ineffective in protecting the virus removal filter. Insoluble dimeric aggregates can form from non-native product molecules. These aggregates can be detected by SEC. However reversible aggregates can form from the native mAb. The formation of these reversible aggregates is concentration dependent; dilution of the feed will cause them to revert to the monomeric state; thus, they cannot be detected by SEC. However, DLS measurements can provide at least some indication of the presence of these reversible aggregates.

Membrane adsorbers can be used to disrupt the reversible aggregates that form and hence arrest the flux decline during virus removal filtration. It is important that the reversible aggregates interact with the ligands present in the membrane adsorber. HIC membrane adsorbers were found to be an effective pretreatment step over a range of feed pH values. Ion exchange membrane adsorbers can also reduce the fouling during virus filtration, but their effectiveness is much more pH dependent. For ion exchange adsorbers, significant improvement in virus filter performance was only obtained when the mAb and adsorber ligands are oppositely charged.

Author Contributions: S.I. Investigation, Writing—Original Draft Preparation; S.-T.C. Software, Data Curation: R.D. Investigation; H.S. Writing—Review and Editing; D.S., Writing—Review and Editing; A.L.Z. Supervision, Funding Acquisition, Writing—Review and Editing, X.Q. Methodology, Resources, Supervision, Funding Acquisition, Writing—Review and Editing, S.R.W. Resources, Supervision, Funding Acquisition, Writing—Original Draft Preparation, Writing—Review and Editing. All authors have read and agreed to the published version of the manuscript.

Funding: Funding for this work was provided by the National Science Foundation Industry and University Cooperative Research Center for Membrane Applications Science and Technology (MAST) via grant numbers 1822101 and 1841474.

Institutional Review Board Statement: Not applicable.

Informed Consent Statement: Not applicable.

Data Availability Statement: The original contributions presented in this study are included in the article. Further inquiries can be directed to the corresponding authors.

Acknowledgments: Negin Nejatishahidein helped with some of the initial experimental studies.

Conflicts of Interest: S.I. was employed by EMD Millipore Corporation. H.S was employed by Asahi Kasei Medical. D.S. was employed by Asahi Kasei Bioprocess America. The remaining authors declare that the research was conducted in the absence of any commercial or financial relationships that could be construed as a potential conflict of interest.

References

1. Mateu, M.G. Mechanical properties of viruses analyzed by atomic force microscopy: A virological perspective. *Virus Res.* **2012**, *168*, 1–22. [CrossRef] [PubMed]
2. Billups, M.; Minervini, M.; Holstein, M.; Feroz, H.; Ranjan, S.; Hung, J.; Bao, H.; Li, Z.J.; Ghose, S.; Zydney, A.L. Role of membrane structure on the filtrate flux during monoclonal antibody filtration through virus retentive membranes. *Biotechnol. Prog.* **2022**, *38*, e3231. [CrossRef] [PubMed]
3. Faqi, A.S. *A Comprehensive Guide to Toxicology in Preclinical Drug Development*; Academic Press: Cambridge, MA, USA, 2012.
4. Wickramasinghe, S.R.; Stump, E.D.; Grzenia, D.L.; Husson, S.M.; Pellegrino, J. Understanding virus filtration membrane performance. *J. Membr. Sci.* **2010**, *365*, 160–169. [CrossRef]
5. Kaufman, Y.; Hunt, K.; Hale, G.; McClure, M.; Latulippe, D.; Sivan, M.; Wilson, J.; Dorin, R.; Agroskin, Y.; Siwak, M. Fouling of virus filtration membranes by monoclonal antibody feeds with low aggregate content. *Biotechnol. Bioeng.* **2024**, *121*, 2400–2408. [CrossRef]
6. Bolton, G.R.; Basha, J.; LaCasse, D.P. Achieving high mass-throughput of therapeutic proteins through parvovirus retentive filters. *Biotechnol. Prog.* **2010**, *26*, 1671–1677. [CrossRef]
7. Brown, A.; Bechtel, C.; Bill, J.; Liu, H.; Liu, J.; McDonald, D.; Pai, S.; Radhamohan, A.; Renslow, R.; Thayer, B. Increasing parvovirus filter throughput of monoclonal antibodies using ion exchange membrane adsorptive pre-filtration. *Biotechnol. Bioeng.* **2010**, *106*, 627–637. [CrossRef]
8. Rayfield, W.J.; Roush, D.J.; Chmielowski, R.A.; Tugcu, N.; Barakat, S.; Cheung, J.K. Prediction of viral filtration performance of monoclonal antibodies based on biophysical properties of feed. *Biotechnol. Prog.* **2015**, *31*, 765–774. [CrossRef] [PubMed]
9. Stanevich, V.; Pachalla, A.; Nunez, B.; McInnes, M.; Nieder, C.; Schreffler, J. Improving viral filtration capacity in biomanufacturing processes using aggregate binding properties of polyamide-6, 6. *Biotechnol. Bioeng.* **2021**, *118*, 1105–1115. [CrossRef]
10. Bolton, G.R.; Spector, S.; LaCasse, D. Increasing the capacity of parvovirus-retentive membranes: Performance of the Viresolve™ Prefilter. *Biotechnol. Appl. Biochem.* **2006**, *43*, 55–63. [CrossRef] [PubMed]
11. Shirataki, H.; Yokoyama, Y.; Oguri, R. Effect of mixed-mode and surface-modified column chromatography on virus filtration performance. *Biochem. Eng. J.* **2021**, *172*, 108034. [CrossRef]
12. Johnson, S.A.; Chen, S.; Bolton, G.; Chen, Q.; Lute, S.; Fisher, J.; Brorson, K. Virus filtration: A review of current and future practices in bioprocessing. *Biotechnol. Bioeng.* **2022**, *119*, 743–761. [CrossRef] [PubMed]
13. Isu, S.; Vinskus, L.; Silva, D.; Cunningham, K.; Elich, T.; Greenhalgh, P.; Sokolnicki, A.; Raghunath, B. Leveraging bioanalytical characterization of fractionated monoclonal antibody pools to identify aggregation-prone and less filterable proteoforms during virus filtration. *Biotechnol. Prog.* **2024**, *40*, e3451. [CrossRef]
14. Barnard, J.G.; Singh, S.; Randolph, T.W.; Carpenter, J.F. Subvisible particle counting provides a sensitive method of detecting and quantifying aggregation of monoclonal antibody caused by freeze-thawing: Insights into the roles of particles in the protein aggregation pathway. *J. Pharm. Sci.* **2011**, *100*, 492–503. [CrossRef]
15. Bieberbach, M.; Kosiol, P.; Seay, A.; Bennecke, M.; Hansmann, B.; Hepbildikler, S.; Thom, V. Investigation of fouling mechanisms of virus filters during the filtration of protein solutions using a high throughput filtration screening device. *Biotechnol. Prog.* **2019**, *35*, e2776. [CrossRef] [PubMed]
16. Arzenšek, D.; Kuzman, D.; Podgornik, R. Colloidal interactions between monoclonal antibodies in aqueous solutions. *J. Colloid Interface Sci.* **2012**, *384*, 207–216. [CrossRef]
17. Esfandiary, R.; Hayes, D.B.; Parupudi, A.; Casas-Finet, J.; Bai, S.; Samra, H.S.; Shah, A.U.; Sathish, H.A. A systematic multitechnique approach for detection and characterization of reversible self-association during formulation development of therapeutic antibodies. *J. Pharm. Sci.* **2013**, *102*, 3089–3099. [CrossRef] [PubMed]
18. Arosio, P.; Rima, S.; Morbidelli, M. Aggregation mechanism of an IgG2 and two IgG1 monoclonal antibodies at low pH: From oligomers to larger aggregates. *Pharm. Res.* **2013**, *30*, 641–654. [CrossRef] [PubMed]
19. Philo, J.S. Is any measurement method optimal for all aggregate sizes and types? *AAPS J.* **2006**, *8*, E564–E571. [CrossRef]
20. Roberts, D.; Keeling, R.; Tracka, M.; Van Der Walle, C.; Uddin, S.; Warwicker, J.; Curtis, R. Specific ion and buffer effects on protein–protein interactions of a monoclonal antibody. *Mol. Pharm.* **2015**, *12*, 179–193. [CrossRef] [PubMed]
21. Zhang, T.; Xu, J.; Chen, J.; Wang, Z.; Wang, X.; Zhong, J. Protein nanoparticles for Pickering emulsions: A comprehensive review on their shapes, preparation methods, and modification methods. *Trends Food Sci. Technol.* **2021**, *113*, 26–41. [CrossRef]
22. Wang, M.; Liu, S.; Ji, H.; Yang, T.; Qian, T.; Yan, C. Salting-out effect promoting highly efficient ambient ammonia synthesis. *Nat. Commun.* **2021**, *12*, 3198. [CrossRef] [PubMed]

23. Philo, J.S.; Arakawa, T. Mechanisms of protein aggregation. *Curr. Pharm. Biotechnol.* **2009**, *10*, 348–351. [CrossRef]
24. Yadav, S.; Laue, T.M.; Kalonia, D.S.; Singh, S.N.; Shire, S.J. The influence of charge distribution on self-association and viscosity behavior of monoclonal antibody solutions. *Mol. Pharm.* **2012**, *9*, 791–802. [CrossRef] [PubMed]
25. Esfandiary, R.; Parupudi, A.; Casas-Finet, J.; Gadre, D.; Sathish, H. Mechanism of reversible self-association of a monoclonal antibody: Role of electrostatic and hydrophobic interactions. *J. Pharm. Sci.* **2015**, *104*, 577–586. [CrossRef] [PubMed]
26. Lohman, T.M.; Overman, L.B.; Ferrari, M.E.; Kozlov, A.G. A highly salt-dependent enthalpy change for escherichia coli SSB Protein− Nucleic acid binding due to Ion− Protein interactions. *Biochemistry* **1996**, *35*, 5272–5279. [CrossRef]
27. Jones, M.; Palackal, N.; Wang, F.; Gaza-Bulseco, G.; Hurkmans, K.; Zhao, Y.; Chitikila, C.; Clavier, S.; Liu, S.; Menesale, E. "High-risk" host cell proteins (HCPs): A multi-company collaborative view. *Biotechnol. Bioeng.* **2021**, *118*, 2870–2885. [CrossRef] [PubMed]

membranes

MDPI

Communication

A Novel Method for Separating Full and Empty Adeno-Associated Viral Capsids Using Ultrafiltration

Deepraj Sarmah and Scott M. Husson *

Department of Chemical and Biomolecular Engineering, Clemson University, 127 Earle Hall, Clemson, SC 29634, USA; dsarmah@clemson.edu
* Correspondence: shusson@clemson.edu; Tel.: +1-(864)-656-4502

Abstract: Adeno-associated viral vectors (AAVs) are the predominant viral vectors used for gene therapy applications. A significant challenge in obtaining effective doses is removing non-therapeutic empty viral capsids lacking DNA cargo. Current methods for separating full (gene-containing) and empty capsids are challenging to scale, produce low product yields, are slow, and are difficult to operationalize for continuous biomanufacturing. This communication demonstrates the feasibility of separating full and empty capsids by ultrafiltration. Separation performance was quantified by measuring the sieving coefficients for full and empty capsids using ELISA, qPCR, and an infectivity assay based on the live cell imaging of green fluorescent protein expression. We demonstrated that polycarbonate track-etched membranes with a pore size of 30 nm selectively permeated empty capsids to full capsids, with a high recovery yield (89%) for full capsids. The average sieving coefficients of full and empty capsids obtained through ELISA/qPCR were calculated as 0.25 and 0.49, indicating that empty capsids were about twice as permeable as full capsids. Establishing ultrafiltration as a viable unit operation for separating full and empty AAV capsids has implications for developing the scale-free continuous purification of AAVs.

Keywords: AAV purification; AAV ultrafiltration; AAV membrane filtration; AAV PCTE membrane; AAV infectivity; AAV live cell imaging; AAV ELISA/qPCR; AAV TEM

1. Introduction

Recombinant adeno-associated viral vectors (AAVs) are fast emerging as the predominant delivery vectors for gene therapies due to their low immunogenicity, non-pathogenicity to humans, and long-term gene expression [1–3]. The presence of multiple AAV serotypes responsive to specific cell types enables targeted gene therapies that improve treatment outcomes [2,3]. Recent U.S. Food and Drug Administration Agency (FDA) approvals of AAV-based treatments (Luxturna® for RPE65 mutation-associated retinal dystrophy [4–6] in 2017; Zolgensma® for Spinal Muscular Atrophy [7] in 2019; Hemgenix® for Hemophilia B [8] in 2022; and Roctavian® for Hemophilia A [9] and Elevedys® for Duchenne muscular dystrophy [10,11] in 2023) indicate the increasing use and diversity of potential AAV treatments. AAV-based gene therapies are being investigated in at least 200 clinical trials [2,3] for conditions such as neurological, hematological, muscular, and ocular diseases and disorders.

There are critical challenges in producing AAVs to meet commercial and clinical demands while maintaining their affordability [1,12]. The high production costs due to low industrial product yields relative to clinical demand contribute to the high costs of gene therapy and limit the broad applicability of AAV-based gene therapies [1,12,13]. To meet this demand, manufacturers have pursued optimizing the AAV capsid yield for several platforms, including using plasmid expression in human embryonic kidney (HEK293) and Chinese hamster ovary (CHO) cell suspension cultures and baculovirus expression vectors in SF9 insect cells [14,15]. Another production challenge is the presence of capsids that are ineffectively packed with DNA cargo or empty [16,17]. Recombinant AAVs can have a lower packing efficiency than wild-type capsids [18], and a higher lot-to-lot variability of the final product has been observed [16,17]. Empty capsids may account for as much as 90% of all capsids [19], and manufacturers must report the purity of the final product according to FDA guidelines [20]. Besides being an impurity in the final product, empty capsids can competitively inhibit full AAVs [19,21], reducing the product's efficacy.

Selectively removing the empty capsids produced during capsid assembly without a loss of full capsids is one of the significant challenges to overcome when attempting to improve product yield [12]. Full and empty capsids differ slightly in their mass and charge, which can be exploited by gradient ultracentrifugation, anion exchange, or multimodal metal affinity chromatography. However, the traditional method of using gradient ultracentrifugation to remove empty capsids [22] is challenging to scale and validate when producing large doses of enriched full AAV capsids [23]. Approaches based on anion exchange or multimodal metal affinity chromatography [23–25] produce a low AAV capsid yield [25], are slow, and are difficult to operationalize for continuous biomanufacturing. Therefore, there is a pressing need for scale-free separation techniques that achieve high product yields.

Although empty capsids are considered to be impurities and inhibitors of AAV therapies, some studies show they may have therapeutic benefits as decoys to overcome AAV clearance [26–28]. Empty capsids bind to neutralizing antibodies produced by the immune system, partially shielding full AAV capsids from the immune response, thereby allowing more functional AAV capsids to reach the target cells. Separation techniques to enrich empty capsids can help to tune the quantity and ratio of full and empty capsids for an effective therapeutic dose.

Separation processes to purify full capsids require a reliable quantification of the full and empty capsids in a final product [25,29–31]. A compilation of prevalent quantification methods [29] demonstrates that obtaining the content percentage and titer of the total capsids or genome (implying full capsids) can be based on a single measurement (e.g., optical density and size exclusion chromatography with multi-angle light scattering) or separate measurements of the total capsid titer (e.g., enzyme-linked immunosorbent assay (ELISA), bio-layer interferometry (BLI), flow virometry (FV), static light scattering combined with dynamic light scattering (SLS-DLS)) and genome titer (e.g., quantitative polymerase chain reaction (qPCR), digital droplet PCR (ddPCR), and dye-based binding assays (Dye-BA)). Other methods can provide content ratio information but not the total capsid or genome titer (e.g., transmission electron microscopy with negative stain (TEM), cryo-electron microscopy (cryo-EM), anion-exchange chromatography (AEC), and charge-detection mass spectrometry (CDMS)). However, only the ELISA, qPCR, ddPCR, FV, BLI, and Dye-BA methods are applicable at low concentrations (below 10^{12} capsids/mL). Among these, serotype-specific ELISA, qPCR, and ddPCR are the most used techniques to quantify capsids. BLI is a relatively new technique based on the interference pattern of white light reflected from a layer of immobilized proteins and rivals ELISA with a potentially larger dynamic range. However, measurements require an Octet optical biosensor. FV and Dye-BA may be good options once their robustness and accuracy become well-established.

PCR techniques to quantify the AAV genome may rely on sequences within inverted terminal repeats (ITRs), regulatory elements, or a transgene. Since ITRs are conserved among different types of AAVs, this is considered to be a more universal method. Comparing qPCR and ddPCR, the latter has a smaller coefficient of variation. However, Wang et al. showed that, by using AAVs of known concentrations as standards, the coefficient of variation in qPCR might be comparable to that of ddPCR [32,33].

The reported vector genome may not necessarily report vector potency because of the possibility of non-functional DNA fragments packed in AAVs. Capsid damage may also occur during purification processes due to mechanical or chemical stresses. Therefore, it is essential to measure AAV functionality. Infectivity assays based on transgene expression measured through flow cytometry, plaque detection, or endpoint assays have been used to quantify the relative functionality of AAVs. Live cell imaging is a powerful tool for measuring transgene expression, involving fewer manual steps than flow cytometry and comparable standard deviations for both lentiviruses [34,35] and AAVs [36].

Our approach looks at viral capsid separation using ultrafiltration. Although ultrafiltration has been used extensively for clarifying and concentrating AAVs and other viruses [37–40], no studies have explored using ultrafiltration for separating empty and full capsids. AAV capsids are speculated to deform under compression to pass through cellular nuclear pores [41], and empty capsids have been shown to deform into ellipsoid shapes through atomic force microscopy [41,42] and selectively pass through nano-pores due to nanopore-induced electro-deformation [43]. Selective permeation through membrane pores based on deformation has been reported to separate supercoiled, linear, and open-circular plasmid isoforms of DNA [44–46]. We theorize that the same principle could separate full and empty capsids. Establishing ultrafiltration as a potentially viable unit operation for separating full and empty AAV capsids has implications for developing the scale-free continuous purification of AAV-based gene therapies.

2. Materials and Methods

2.1. Virus Source and Storage

We procured recombinant AAV2 enriched in full and empty capsids from Virovek, Inc. (Houston, TX, USA) [47]. The full-enriched capsids (product AAV-CMV-GFP) can express green fluorescent protein (GFP) after transduction in mammalian cells. The plasmids used to create empty capsids (product AAV-empty) did not contain any AAV inverted terminal repeats flanking the DNA to be packed into the AAV. The capsid concentrations for full and empty capsid products were reported to be $>10^{13}$ capsids/mL, and we confirmed this through assays (see the following sections). The solution was divided into 10 μL aliquots and stored at −80 °C. Storage in 10 μL aliquots prevented the products from undergoing multiple freeze–thaw cycles. We did not observe any significant changes in the concentrations of the viral capsids after 8 months of cold storage.

2.2. Selection of Buffer

Factors related to the stability of AAVs in a buffer, like ionic strength, pH, and the presence of surfactants and free radical oxidation inhibitors, were considered when choosing the working buffer. The aggregation of AAVs is known to be high in pure water, but it decreases and plateaus with an increased ionic strength [48]. The stability of AAVs and preventing adsorption onto surfaces are enhanced by non-ionic detergents like poloxamer 188 (Pluronic F-68) and polysorbate 80 [49]. The presence of ethylenediaminetetraacetic acid (EDTA) mitigates the effects of nucleases, often found at the end of manufacturing processes [49]. AAVs are known to be stable from pH 5.5 to 8.5 and have isoelectric points of 5.9 for full capsids and 6.3 for empty capsids [50].

Our buffer used phosphate-buffered saline (PBS, Fisher Scientific, Waltham, MA, USA) with an ionic strength of 163 mM and 0.1% Pluronic F-68 (Gibco, Waltham, MA, USA) to prevent adsorption onto surfaces. Nine millimolar ascorbic acid (Fisher Scientific) was used to scavenge free radicals and decrease the pH to about 6.0 ± 0.1. One millimolar EDTA (Fisher Scientific) was used to protect the DNA from the effect of nucleases. Our experiments always used a freshly prepared buffer to avoid ascorbic acid degradation.

2.3. Determination of Assays

Quantification methods for AAVs are numerous [29–31], but many require a relatively high concentration of capsids ($>10^{12}$ capsids/mL). Since dilution in a buffer results in a viral capsid concentration of ~10^{10}, our assay options were limited. Figure 1 illustrates the three principal assays used to quantify the total and full/infectious titers of the AAV solutions before and after the separation steps.

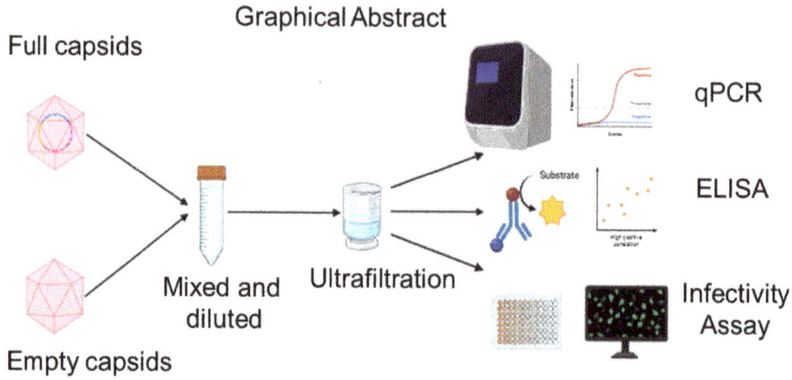

Figure 1. Graphical abstract of experimental protocols. Ultrafiltration of a mixture of full and empty capsids was followed by analysis through qPCR (quantitative polymerase chain reaction), ELISA (enzyme-linked immunosorbent assay), and infectivity assay.

2.3.1. ELISA

We used the AAV2-specific ELISA kit (catalog number: PRAAV2R) from Progen Biotechnik (Heidelberg, Germany) to quantify the concentrations of total capsids. The final measurements were made with a Synergy H1 microplate reader (Aligent BioTek, Winooski, VA, USA), and the results were obtained by comparing them with a linear fit of absorbance at 450 nm plotted against concentration.

2.3.2. qPCR

We used qPCR to measure the vector genome number. The primer (sourced from Integrated DNA Technologies, Coralville, IA, USA) was based on ITR sequences (forward: 5-GGA ACC CCT AGT GAT GGA GTT-3; reverse: 5-CGG CCT CAG TGA GCG A-3). AAVs of known concentrations were used as standards (Addgene, Watertown, MA, USA; catalog number 59462-AAV2). Earlier studies established that this approach has fewer steps and less uncertainty than using DNA standards [32,33]. To increase precision, the samples were loaded with an Opentrons® OT-2 automated pipette (Opentrons Labworks, Inc., Long Island City, NY, USA). Real-time measurements were obtained through a qPCR cycler and Bio-Rad CFX Connect (Aligent BioTek) detector.

2.3.3. Infectivity Assay

Parallel to measuring the vector genome, we developed live cell imaging based on the expression of GFP to measure the functional capsids. HEK293T cells (Addgene) were seeded overnight in a 96-well plate, with 150 μL of full-growth DMEM 1Xmedia (Gibco) and 4500 cells per well. Six concentrations were generated through serial half dilutions, starting with 1 mL of an AAV solution with a known vector genome concentration of ~10^{10} capsids/mL (determined by qPCR). Ten microliters of each dilution were added to the wells with the HEK293T cells to generate four replicate wells for each AAV concentration. After 48 h, phase-contrast and fluorescent images were taken of each well using a Cytation 5 imager (Aligent BioTek), keeping the exposure settings constant across the wells. Using the open-source software CellProfiler (v4.2.6 and v4.2.7) [51,52], the fluorescence area and intensity were measured in each well and divided by the corresponding total cytoplasm area. These values were plotted against concentration to generate calibration curves fitted to a power law model.

2.3.4. TEM

Finally, we observed and generated images of the AAV-CMV-GFP (full-enriched) and AAV-empty (empty-enriched) products from Virovek using TEM. A droplet (3 μL) of AAV product was placed on a carbon film-coated TEM grid (Ted Pella, Redding, CA, USA) for 30 s, followed by three rinse steps with 3 µL of distilled water and staining in an aqueous solution of 1% (w/w) uranyl acetate for 30 s. Imaging was performed at 30,000× in a TEM microscope (HT7830, Hitachi, Tokyo, Japan). The TEM images were analyzed using the CellProfiler software to quantify the full/empty capsid ratio for each product.

2.4. Dilution to Desired Ratio and Concentration

Solutions were prepared by mixing 10 µL of full capsid solution and enough empty capsid solution to ensure a 50/50 ratio of full to empty capsids, then diluting the mixture to 25 mL with the buffer. This ensured that the concentration of the capsids after dilution was high enough to be measurable by ELISA. This concentration was also high enough for mammalian cells to express the gene of interest at quantifiable levels for fluorescent imaging.

2.5. Ultrafiltration

Ultrafiltration experiments were performed with a 50 mL Amicon® stirred cell (UFSC05001) from MilliporeSigma (Burlington, MA, USA). Commercial membranes were selected with pore sizes close to the AAV capsid diameter measured by TEM. We tested 25 nm (VSWP04700) mixed cellulose ester (MCE) membranes from MilliporeSigma and 30 nm (1270011), 50 nm (PCT0059030), and 80 nm (PCT0089030) hydrophilic polycarbonate track-etched (PCTE) membranes from Sterlitech (Auburn, WA, USA). We selected the MCE membrane for its precise pore size. We included the hydrophilic PCTE membranes due to their low pore size distribution and low protein binding. This range of pore sizes was established in preliminary experiments that found no capsid permeation using the 25 nm membrane and no separation using the 80 nm membrane. Testing commercial membranes made from other polymers was beyond the scope of this feasibility study. We performed ultrafiltration at a pressure of 69 kPa for all the membranes except the 30 nm PCTE, which had a relatively low flow rate, for which we increased the pressure to 172 kPa.

One milliliter of the 25 mL feed AAV solution was collected for concentration measurements. The remaining 24 mL was loaded into the stirred cell, and the magnetic stirrer was set to 300 RPM. The cell was pressurized, and ultrafiltration was performed for a time sufficient to collect 12 mL of permeate. Samples of the permeate and retentate were collected for concentration measurements. We measured the total capsid concentrations of the original feed, permeate, and retentate by ELISA, using the necessary dilutions to bring the concentrations to within the dynamic range of the kit. qPCR was performed for the same diluted samples. A mass balance was used to determine the empty capsid concentration by taking the difference between the total and full capsid concentrations.

3. Results

3.1. AAV2 Capsids Are Fully Retained by the 25 nm MCE Membrane and Pass Freely through 50 nm and 80 nm PCTE Membranes

Using ELISA, we measured the total capsid concentrations of the full- and empty-enriched capsid solutions from Virovek to be 2.9×10^{13} and 6.3×10^{13} total capsids/mL. From qPCR, the vector genome concentration of the full-enriched capsid solution was 2.7×10^{13}, indicating 93% full capsids. The vector genome content of the empty-enriched capsid solution was insignificant (i.e., the concentration of DNA with ITR was <0.01% than that of the full-enriched solution). These data were used to prepare the feed solutions with a 50/50 ratio of full to empty capsids.

Figure 2a illustrates a hypothetical example of perfect separation, in which only empty capsids pass through a membrane. Half the original feed is passed through a membrane in this hypothetical example. The empty capsids pass freely through the membrane, and the full capsids are rejected completely. In this hypothetical example, the permeate contains half the empty capsids, and the retentate contains half the empty capsids and all the full capsids. In the case of no separation (Figure 2b), the full and empty capsids pass freely through the membrane, and there is no difference between the permeate and retentate. Each would contain half the full and empty capsids.

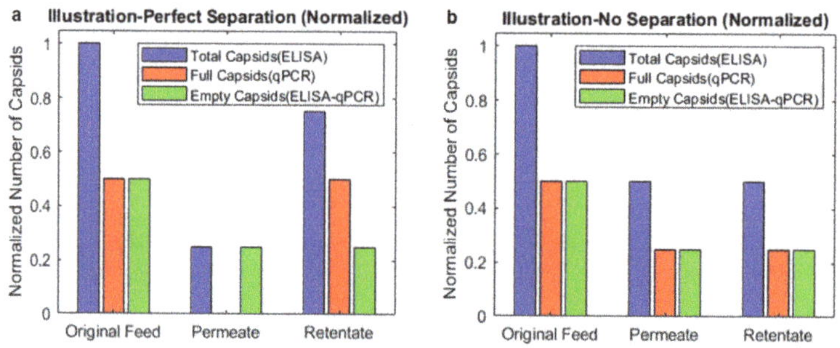

Figure 2. Illustration of hypothetical cases of perfect separation and no separation when half of the feed is passed through the membrane. For perfect separation, all the full capsids are retained in the retentate, and the permeate contains only empty capsids. For no separation, both full and empty capsids are completely permeable, and the permeate and retentate end up having the same composition as the original feed.

For the 25 nm MCE membrane, the empty and full capsids were fully retained. No capsids were detected in the permeate by ELISA, and the qPCR signal was negligible (<0.1% compared to the original feed). This indicated that we should test membranes with larger pores to enable AAV capsids to pass through the membrane. However, experiments using the PCTE membranes with 50 nm and 80 nm pores showed no separation of full and empty capsids (Figure 3a,b). The recoveries were greater than 90% for the total, full, and empty capsids based on mass balances.

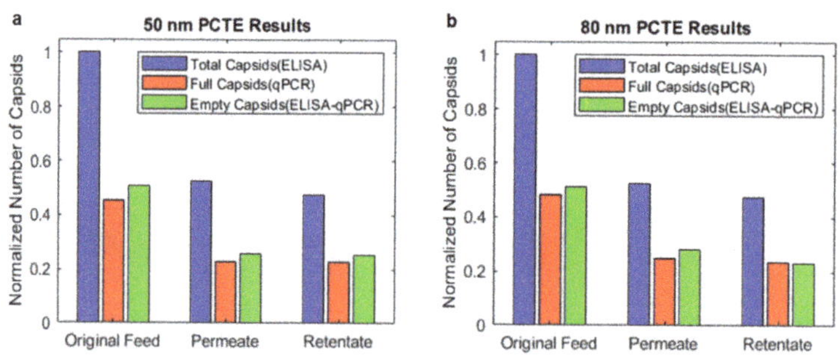

Figure 3. Results from (**a**) 50 nm and (**b**) 80 nm PCTE ultrafiltration experiments. No significant separation of full and empty capsids was observed.

3.2. Combining ELISA and qPCR Results Indicates the Enrichment of Empty Capsids in the Permeate and Full Capsids in the Retentate for 30 nm PCTE Membranes

For experiments with the 30 nm PCTE membrane, we observed that the ratio of full capsids in the retentate to the original feed, measured by qPCR, was higher than that of the total capsids, measured by ELISA. This shows an enrichment of full capsids in the retentate. The reverse was true for the permeate (i.e., the ratio of total capsids in the permeate to the original feed was higher than the ratio of full capsids), indicating the enrichment of empty capsids in the permeate.

We combined the results of the ELISA and qPCR to generate Figure 4a. We observed that adding empty-enriched capsids to full-enriched capsids (see Section 2) generally resulted in an original feed that was slightly higher in full capsids, on average, than the expected 50/50 split of full and empty capsids. However, the ratio of full capsids/total capsids in the retentate was always greater than that in the original feed. The permeate had a lower ratio of full capsids/total capsids than the original feed. This shows that 30 nm PCTE membranes are selectively permeable to AAV empty capsids and that AAV capsids may be enriched through ultrafiltration using this membrane.

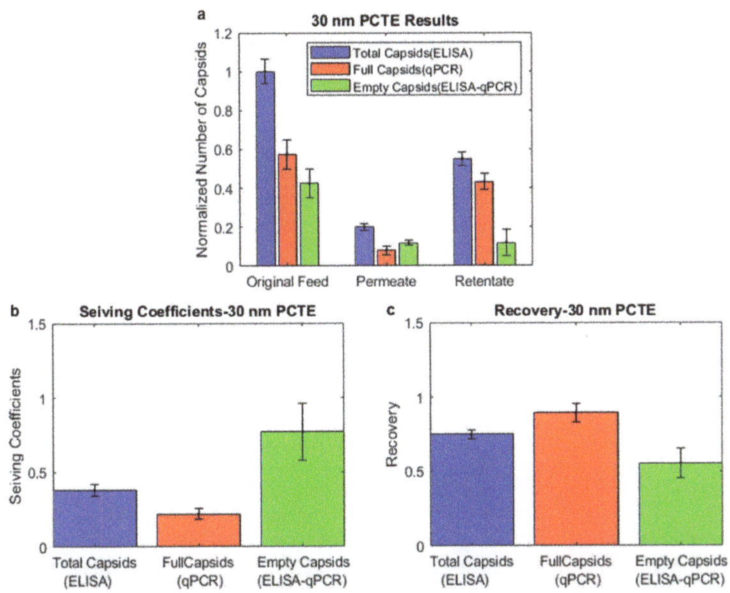

Figure 4. Results from 30 nm PCTE ultrafiltration experiments and analysis through ELISA/qPCR. (**a**) Enrichment of full capsids in the retentate and empty capsids in the permeate were observed, indicating higher permeability of empty capsids. (**b**) Analysis of the sieving coefficients shows that the sieving coefficient of empty capsids is higher than full capsids. (**c**) Analysis of the recoveries shows that the recovery of full capsids is higher than empty capsids.

To analyze these results further, we compared the sieving coefficients of the total, full, and empty capsids. An average sieving coefficient for each experiment was obtained by dividing the permeate concentrations of the total, full, and empty capsids with the averages of the initial concentrations (original feed) and final retentate concentrations. The average sieving coefficients of the total, full, and empty capsids were calculated as 0.38, 0.22, and 0.77, indicating that empty capsids may be about 3.5 times more permeable than full capsids (Figure 4b). The recoveries of the total, full, and empty capsids were calculated as 0.75, 0.89, and 0.55, indicating that more empty capsids than full capsids were lost during the ultrafiltration process.

3.3. Combining ELISA and Infectivity Assays also Indicates Enrichment of Empty Capsids in the Permeate and Full Capsids in the Retentate for 30 nm PCTE Membranes

We further combined the ELISA and infectivity assay results to validate the enrichment of full capsids in the retentate (Figure 5a). The percentage of full capsids in the original feed was slightly below 50% for the infectivity experiments. However, after ultrafiltration, the proportion of full capsids in the retentate compared to the original feed was always greater than the proportion of total capsids. This validated that the 30 nm PCTE membranes were preferably permeable to empty capsids.

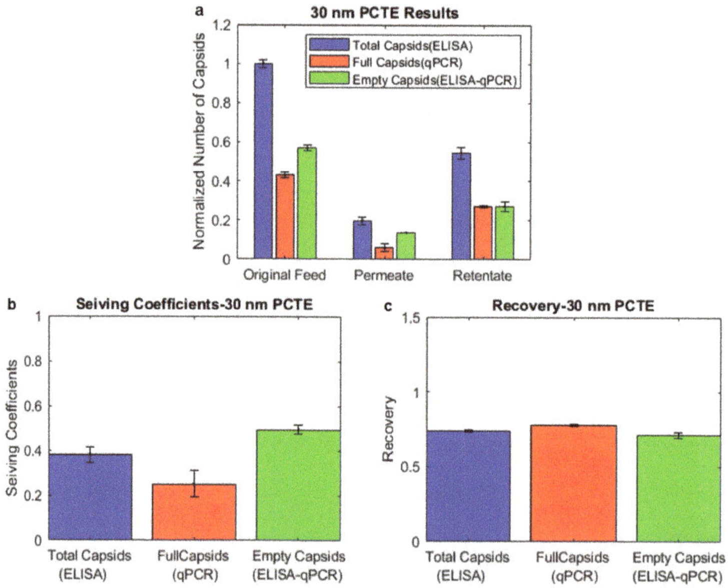

Figure 5. Results from 30 nm PCTE ultrafiltration experiments and analysis through ELISA/infectivity assays. (**a**) Enrichment of full capsids in the retentate and empty capsids in the permeate was observed, indicating higher permeability of empty capsids. (**b**) Analysis of the sieving coefficients shows that the sieving coefficient of empty capsids is higher than full capsids. (**c**) Analysis of the recoveries shows that the recovery of full capsids is marginally higher than that of empty capsids.

The average sieving coefficients of the total, full, and empty capsids obtained through the ELISA/infectivity assays were calculated as 0.38, 0.25, and 0.49, indicating that empty capsids may be about twice as permeable as full capsids. The enrichment of full capsids in the retentate was lower than that measured by ELISA/qPCR. The possible reasons for this are discussed in Section 4. The recoveries of the total, full, and empty capsids were calculated as 0.74, 0.78, and 0.72, indicating that marginally more empty capsids were lost during the ultrafiltration process.

4. Discussion

Several sources manufacture AAVs for laboratory use, including Virovek, Charles River, and Sirion Biotek. Due to their ease of procurement and cost, we used Virovek products derived from an insect-cell-based production process. These insect-cell-based products were initially thought to be indistinguishable from those made using mammalian cells [53]. Ebbernick et al. highlighted some properties of the insect-cell-based AAVs from Virovek compared to other manufacturers. They found differences in the distribution of full versus empty capsids and saw that AAVs produced with insect cells contained a higher proportion of overloaded capsids [54]. Furthermore, the full AAV-GFP-CMV capsids have a vector genome size of ~2.3 kDa, which is lower than the maximum of ~4.7 kDa possible for AAVs. It is possible that a full capsid could pack more than one 2.3 kDa vector genome, which could have implications for the interpretation of assay results.

Supposing that overloaded AAV capsids contain more than one pair of ITR sequences, in this case, we can theorize that, during the separation of full and empty capsids, overloaded capsids and full capsids will be enriched in the retentate. This would cause the qPCR results for the retentate to produce a higher value than expected relative to the original capsids. This could explain the differences between the ELISA/qPCR and ELISA/infectivity assay results, where the ratio of full capsids/total capsids was relatively higher in the qPCR results than in the infectivity assay.

Precautions were taken when formulating the buffer to prevent the adsorption of AAV particles onto the surfaces of the ultrafiltration apparatus and membrane. Nevertheless, we observed higher recoveries (>90%) for the total, full, and empty capsids for the 50 nm and 80 nm PCTE membranes than the 30 nm PCTE membrane. It appeared that some AAV particles were retained within the pores of the membranes, and this effect may have been more significant for the 30 nm membranes due to their similar pore size to the AAVs. Since recovery was higher for the full capsids than the empty capsids, empty capsids may have been more prone to be retained within the membrane pores, which was expected, since a higher proportion of empty capsids entered the pores.

The purity of full capsids after separation by ultracentrifugation can range from 70% to 95%, depending on the precision of the salt gradient and the ability to accurately fractionate the gradient, and the yield of full capsids typically ranges from 50% to 80% [40,55–57]. Chromatography-based methods can achieve purities ranging from 80% to 95% for full AAV capsids, and the recovery of full capsids typically ranged from 50% to 90% in some recent studies [23,24,57–60]. Our recovery measured through qPCR was 89% for full capsids, and the purity increased from an average of 59% to 82% full capsids after one pass. Unlike ultracentrifugation and chromatography, our ultrafiltration method is easy to operationalize for continuous bioprocessing. Unlike antibody-based chromatography methods, our ultrafiltration method can potentially be used across serotypes because of the common size of capsids.

The mechanism of selective permeation is thought to be differences in deformation between empty and full capsids, which has been observed in AFM studies [41,42] and nanopore-induced electro-deformation studies [43]. We theorize that a higher deformation of empty capsids during pressure-driven filtration caused them to have a larger sieving coefficient. Two more potentially exploitable differences between full and empty capsids are differences in hydrodynamic diameter and differences in affinity to certain materials. The reported hydrodynamic diameters of full and empty capsids range from 25 to 40 nm, depending on the measurement method [58,60,61]. Factors like the ionic strength and pH of the buffer may influence these values. Knowing the factors influencing the hydrodynamic diameter may provide insights for size-based separations. Affinity differences between the full/empty capsids and the membrane may enhance separation. Here, again, ionic strength and the presence of Mg^{+2} enhance the affinity differences of full and empty capsids in anion exchange chromatography [20,23,24]. Finally, commercial membranes with a higher porosity than the PCTE membrane used in this feasibility study will increase productivity. This could benefit industrial applications, albeit with the downside that their higher pore size distribution may reduce separation efficiency.

In summary, we report a method for the selective enrichment of full and empty AAV2 capsids based on ultrafiltration. As observed from the infectivity assays, the AAVs retained infectivity after the ultrafiltration process. This makes the method suitable for ultrafiltration–diafiltration strategies for further enrichment and tangential flow filtration (TFF) for processing larger volumes. Transitioning from a stirred cell ultrafiltration setup to a TFF setup is feasible by accounting for differences in flow dynamics, pressure control, and system complexity. Future work is needed to understand the process using one of the many commercial research-scale TFF systems. These TFF systems are highly scalable from the laboratory to pilot to industrial scale.

Author Contributions: Conceptualization, D.S. and S.M.H.; methodology, D.S. and S.M.H.; validation, D.S. and S.M.H.; formal analysis, D.S. and S.M.H.; investigation, D.S.; resources, S.M.H.; data curation, D.S.; writing—original draft preparation, D.S.; writing—review and editing, S.M.H.; visualization, D.S.; supervision, S.M.H.; project administration, S.M.H.; funding acquisition, S.M.H. All authors have read and agreed to the published version of the manuscript.

Funding: This research was funded by the U.S. National Science Foundation, Grant Number OIA-2218054.

Institutional Review Board Statement: The study was approved by the Institutional Biosafety Committee of Clemson University (IBC2023-0102, approved from 2 February 2023 to 1 February 2026) for studies involving recombinant AAV.

Data Availability Statement: The code and data required to generate the figures can be obtained from https://github.com/deeprajs/membranes, accessed on 29 August 2024. Additional data may be provided on request.

Acknowledgments: We acknowledge the support and training provided by Clemson University's electron microscopy facility, especially Dayton Cash (planning and oversight), Haijun Qian (training on analysis of biological samples), and Donald Mulwee (operation of the microscope). S.M.H. acknowledges support from the William B. "Bill" Sturgis, '57 and Martha Elizabeth "Martha Beth" Blackmon Sturgis Distinguished Professorship in Chemical and Biomolecular Engineering.

Conflicts of Interest: The funders had no role in the design of the study; in the collection, analyses, or interpretation of data; in the writing of the manuscript; or in the decision to publish the results.

References

1. Wang, D.; Tai, P.W.L.; Gao, G. Adeno-Associated Virus Vector as a Platform for Gene Therapy Delivery. *Nat. Rev. Drug Discov.* **2019**, *18*, 358–378. [CrossRef] [PubMed]
2. Kuzmin, D.A.; Shutova, M.V.; Johnston, N.R.; Smith, O.P.; Fedorin, V.V.; Kukushkin, Y.S.; van der Loo, J.C.M.; Johnstone, E.C. The Clinical Landscape for AAV Gene Therapies. *Nat. Rev. Drug Discov.* **2021**, *20*, 173–174. [CrossRef] [PubMed]
3. Chancellor, D.; Barrett, D.; Nguyen-Jatkoe, L.; Millington, S.; Eckhardt, F. The State of Cell and Gene Therapy in 2023. *Mol. Ther.* **2023**, *31*, 3376–3388. [CrossRef] [PubMed]
4. Gao, J.; Hussain, R.M.; Weng, C.Y. Voretigene Neparvovec in Retinal Diseases: A Review of the Current Clinical Evidence. *Clin. Ophthalmol.* **2020**, *14*, 3855–3869. [CrossRef] [PubMed]
5. Ameri, H. Prospect of Retinal Gene Therapy Following Commercialization of Voretigene Neparvovec-Rzyl for Retinal Dystrophy Mediated by RPE65 Mutation. *J. Curr. Ophthalmol.* **2018**, *30*, 1–2. [CrossRef]
6. Russell, S.; Bennett, J.; Wellman, J.A.; Chung, D.C.; Yu, Z.-F.; Tillman, A.; Wittes, J.; Pappas, J.; Elci, O.; McCague, S.; et al. Efficacy and Safety of Voretigene Neparvovec (AAV2-hRPE65v2) in Patients with RPE65-Mediated Inherited Retinal Dystrophy: A Randomised, Controlled, Open-Label, Phase 3 Trial. *Lancet* **2017**, *390*, 849–860. [CrossRef]
7. Ogbonmide, T.; Rathore, R.; Rangrej, S.B.; Hutchinson, S.; Lewis, M.; Ojilere, S.; Carvalho, V.; Kelly, I. Gene Therapy for Spinal Muscular Atrophy (SMA): A Review of Current Challenges and Safety Considerations for Onasemnogene Abeparvovec (Zolgensma). *Cureus* **2023**, *15*, e36197. [CrossRef]
8. Tice, J.A.; Walton, S.M.; Sarker, J.; Moradi, A.; Chu, J.N.; Herce-Hagiwara, B.; Fahim, S.M.; Agboola, F.; Rind, D.; Pearson, S.D. The Effectiveness and Value of Gene Therapy for Hemophilia: A Summary from the Institute for Clinical and Economic Review's California Technology Assessment Forum. *J. Manag. Care Spec. Pharm.* **2023**, *29*, 576–581. [CrossRef]
9. Ozelo, M.C.; Mahlangu, J.; Pasi, K.J.; Giermasz, A.; Leavitt, A.D.; Laffan, M.; Symington, E.; Quon, D.V.; Wang, J.-D.; Peerlinck, K.; et al. Valoctocogene Roxaparvovec Gene Therapy for Hemophilia A. *N. Engl. J. Med.* **2022**, *386*, 1013–1025. [CrossRef]
10. Manini, A.; Abati, E.; Nuredini, A.; Corti, S.; Comi, G.P. Adeno-Associated Virus (AAV)-Mediated Gene Therapy for Duchenne Muscular Dystrophy: The Issue of Transgene Persistence. *Front. Neurol.* **2022**, *12*, 814174. [CrossRef]
11. Hoy, S.M. Delandistrogene Moxeparvovec: First Approval. *Drugs* **2023**, *83*, 1323–1329. [CrossRef] [PubMed]
12. Fu, Q.; Polanco, A.; Lee, Y.S.; Yoon, S. Critical Challenges and Advances in Recombinant Adeno-Associated Virus (rAAV) Biomanufacturing. *Biotechnol. Bioeng.* **2023**, *120*, 2601–2621. [CrossRef] [PubMed]
13. Srivastava, A.; Mallela, K.M.G.; Deorkar, N.; Brophy, G. Manufacturing Challenges and Rational Formulation Development for AAV Viral Vectors. *J. Pharm. Sci.* **2021**, *110*, 2609–2624. [CrossRef] [PubMed]
14. Sha, S.; Maloney, A.J.; Katsikis, G.; Nguyen, T.N.T.; Neufeld, C.; Wolfrum, J.; Barone, P.W.; Springs, S.L.; Manalis, S.R.; Sinskey, A.J.; et al. Cellular Pathways of Recombinant Adeno-Associated Virus Production for Gene Therapy. *Biotechnol. Adv.* **2021**, *49*, 107764. [CrossRef] [PubMed]
15. Meier, A.F.; Fraefel, C.; Seyffert, M. The Interplay between Adeno-Associated Virus and Its Helper Viruses. *Viruses* **2020**, *12*, 662. [CrossRef]
16. Schnödt, M.; Büning, H. Improving the Quality of Adeno-Associated Viral Vector Preparations: The Challenge of Product-Related Impurities. *Hum. Gene Ther. Methods* **2017**, *28*, 101–108. [CrossRef]
17. Grieger, J.C.; Soltys, S.M.; Samulski, R.J. Production of Recombinant Adeno-Associated Virus Vectors Using Suspension HEK293 Cells and Continuous Harvest of Vector From the Culture Media for GMP FIX and FLT1 Clinical Vector. *Mol. Ther.* **2016**, *24*, 287–297. [CrossRef]
18. Grimm, D.; Kern, A.; Pawlita, M.; Ferrari, F.K.; Samulski, R.J.; Kleinschmidt, J.A. Titration of AAV-2 Particles via a Novel Capsid ELISA: Packaging of Genomes Can Limit Production of Recombinant AAV-2. *Gene Ther.* **1999**, *6*, 1322–1330. [CrossRef]

19. Gao, K.; Li, M.; Zhong, L.; Su, Q.; Li, J.; Li, S.; He, R.; Zhang, Y.; Hendricks, G.; Wang, J.; et al. Empty Virions in AAV8 Vector Preparations Reduce Transduction Efficiency and May Cause Total Viral Particle Dose-Limiting Side Effects. *Mol. Ther. Methods Clin. Dev.* **2014**, *1*, 9. [CrossRef]
20. Aebischer, M.K.; Gizardin-Fredon, H.; Lardeux, H.; Kochardt, D.; Elger, C.; Haindl, M.; Ruppert, R.; Guillarme, D.; D'Atri, V. Anion-Exchange Chromatography at the Service of Gene Therapy: Baseline Separation of Full/Empty Adeno-Associated Virus Capsids by Screening of Conditions and Step Gradient Elution Mode. *Int. J. Mol. Sci.* **2022**, *23*, 12332. [CrossRef]
21. Urabe, M.; Xin, K.-Q.; Obara, Y.; Nakakura, T.; Mizukami, H.; Kume, A.; Okuda, K.; Ozawa, K. Removal of Empty Capsids from Type 1 Adeno-Associated Virus Vector Stocks by Anion-Exchange Chromatography Potentiates Transgene Expression. *Mol. Ther.* **2006**, *13*, 823–828. [CrossRef] [PubMed]
22. Hermens, W.T.J.M.C.; Brake, O.T.; Dijkhuizen, P.A.; Sonnemans, M.A.F.; Grimm, D.; Kleinschmidt, J.A.; Verhaagen, J. Purification of Recombinant Adeno-Associated Virus by Iodixanol Gradient Ultracentrifugation Allows Rapid and Reproducible Preparation of Vector Stocks for Gene Transfer in the Nervous System. *Human. Gene Ther.* **1999**, *10*, 1885–1891. [CrossRef] [PubMed]
23. Dickerson, R.; Argento, C.; Pieracci, J.; Bakhshayeshi, M. Separating Empty and Full Recombinant Adeno-Associated Virus Particles Using Isocratic Anion Exchange Chromatography. *Biotechnol. J.* **2021**, *16*, e2000015. [CrossRef] [PubMed]
24. Gagnon, P.; Leskovec, M.; Prebil, S.D.; Žigon, R.; Štokelj, M.; Raspor, A.; Peljhan, S.; Štrancar, A. Removal of Empty Capsids from Adeno-Associated Virus Preparations by Multimodal Metal Affinity Chromatography. *J. Chromatogr. A* **2021**, *1649*, 462210. [CrossRef] [PubMed]
25. Chen, H. Manufacturing of Adeno-Associated Viruses, for Example: AAV2. In *Viral Vectors for Gene Therapy: Methods and Protocols*; Merten, O.-W., Al-Rubeai, M., Eds.; Methods in Molecular Biology; Humana Press: Totowa, NJ, USA, 2011; pp. 235–246. ISBN 978-1-61779-095-9.
26. Mingozzi, F.; Anguela, X.M.; Pavani, G.; Chen, Y.; Davidson, R.J.; Hui, D.J.; Yazicioglu, M.; Elkouby, L.; Hinderer, C.J.; Faella, A.; et al. Overcoming Preexisting Humoral Immunity to AAV Using Capsid Decoys. *Sci. Transl. Med.* **2013**, *5*, 194ra92. [CrossRef]
27. Hoffman, B.E.; Herzog, R.W. Covert Warfare Against the Immune System: Decoy Capsids, Stealth Genomes, and Suppressors. *Mol. Ther.* **2013**, *21*, 1648–1650. [CrossRef]
28. Wright, J.F. AAV Empty Capsids: For Better or for Worse? *Mol. Ther.* **2014**, *22*, 1–2. [CrossRef]
29. Gimpel, A.L.; Katsikis, G.; Sha, S.; Maloney, A.J.; Hong, M.S.; Nguyen, T.N.T.; Wolfrum, J.; Springs, S.L.; Sinskey, A.J.; Manalis, S.R.; et al. Analytical Methods for Process and Product Characterization of Recombinant Adeno-Associated Virus-Based Gene Therapies. *Mol. Ther. Methods Clin. Dev.* **2021**, *20*, 740–754. [CrossRef]
30. Werle, A.K.; Powers, T.W.; Zobel, J.F.; Wappelhorst, C.N.; Jarrold, M.F.; Lyktey, N.A.; Sloan, C.D.K.; Wolf, A.J.; Adams-Hall, S.; Baldus, P.; et al. Comparison of Analytical Techniques to Quantitate the Capsid Content of Adeno-Associated Viral Vectors. *Mol. Ther. Methods Clin. Dev.* **2021**, *23*, 254–262. [CrossRef]
31. Dobnik, D.; Kogovšek, P.; Jakomin, T.; Košir, N.; Tušek Žnidarič, M.; Leskovec, M.; Kaminsky, S.M.; Mostrom, J.; Lee, H.; Ravnikar, M. Accurate Quantification and Characterization of Adeno-Associated Viral Vectors. *Front. Microbiol.* **2019**, *10*, 1570. [CrossRef]
32. Wang, Y.; Menon, N.; Shen, S.; Feschenko, M.; Bergelson, S. A qPCR Method for AAV Genome Titer with ddPCR-Level of Accuracy and Precision. *Mol. Ther. Methods Clin. Dev.* **2020**, *19*, 341–346. [CrossRef] [PubMed]
33. Wang, Y.; Cooper, R.; Kiladjian, A.; Bergelson, S.; Feschenko, M. A Digestion-Free Method for Quantification of Residual Host Cell DNA in rAAV Gene Therapy Products. *Mol. Ther. Methods Clin. Dev.* **2019**, *13*, 526–531. [CrossRef] [PubMed]
34. Labisch, J.J.; Wiese, G.P.; Barnes, K.; Bollmann, F.; Pflanz, K. Infectious Titer Determination of Lentiviral Vectors Using a Temporal Immunological Real-Time Imaging Approach. *PLoS ONE* **2021**, *16*, e0254739. [CrossRef]
35. Ye, M.; Keicher, M.; Gentschev, I.; Szalay, A.A. Efficient Selection of Recombinant Fluorescent Vaccinia Virus Strains and Rapid Virus Titer Determination by Using a Multi-Well Plate Imaging System. *Biomedicines* **2021**, *9*, 1032. [CrossRef]
36. Meierrieks, F.; Kour, A.; Pätz, M.; Pflanz, K.; Wolff, M.W.; Pickl, A. Unveiling the Secrets of Adeno-Associated Virus: Novel High-Throughput Approaches for the Quantification of Multiple Serotypes. *Mol. Ther. Methods Clin. Dev.* **2023**, *31*, 101118. [CrossRef] [PubMed]
37. Arunkumar, A.; Singh, N. Ultrafiltration Behavior of Recombinant Adeno Associated Viral Vectors Used in Gene Therapy. *J. Membr. Sci.* **2021**, *620*, 118812. [CrossRef]
38. Singh, N.; Heldt, C.L. Challenges in Downstream Purification of Gene Therapy Viral Vectors. *Curr. Opin. Chem. Eng.* **2022**, *35*, 100780. [CrossRef]
39. Hagen, S.; Baumann, T.; Wagner, H.J.; Morath, V.; Kaufmann, B.; Fischer, A.; Bergmann, S.; Schindler, P.; Arndt, K.M.; Müller, K.M. Modular Adeno-Associated Virus (rAAV) Vectors Used for Cellular Virus-Directed Enzyme Prodrug Therapy. *Sci. Rep.* **2014**, *4*, 3759. [CrossRef]
40. Strobel, B.; Miller, F.D.; Rist, W.; Lamla, T. Comparative Analysis of Cesium Chloride- and Iodixanol-Based Purification of Recombinant Adeno-Associated Viral Vectors for Preclinical Applications. *Human. Gene Ther. Methods* **2015**, *26*, 147–157. [CrossRef]
41. Zeng, C.; Moller-Tank, S.; Asokan, A.; Dragnea, B. Probing the Link among Genomic Cargo, Contact Mechanics, and Nanoindentation in Recombinant Adeno-Associated Virus 2. *J. Phys. Chem. B* **2017**, *121*, 1843–1853. [CrossRef]
42. Zoratto, S.; Weiss, V.U.; Friedbacher, G.; Buengener, C.; Pletzenauer, R.; Foettinger-Vacha, A.; Graninger, M.; Allmaier, G. Adeno-Associated Virus Virus-like Particle Characterization via Orthogonal Methods: Nanoelectrospray Differential Mobility Analysis, Asymmetric Flow Field-Flow Fractionation, and Atomic Force Microscopy. *ACS Omega* **2021**, *6*, 16428–16437. [CrossRef]

43. Karawdeniya, B.I.; Bandara, Y.M.N.D.Y.; Khan, A.I.; Chen, W.T.; Vu, H.-A.; Morshed, A.; Suh, J.; Dutta, P.; Kim, M.J. Adeno-Associated Virus Characterization for Cargo Discrimination through Nanopore Responsiveness. *Nanoscale* **2020**, *12*, 23721–23731. [CrossRef]
44. Borujeni, E.E.; Zydney, A.L. Separation of Plasmid DNA Isoforms Using Centrifugal Ultrafiltration. *BioTechniques* **2012**, *53*, 49–56. [CrossRef]
45. Latulippe, D.R.; Zydney, A.L. Separation of Plasmid DNA Isoforms by Highly Converging Flow through Small Membrane Pores. *J. Colloid Interface Sci.* **2011**, *357*, 548–553. [CrossRef]
46. Li, Y.; Currie, D.; Zydney, A.L. Enhanced Purification of Plasmid DNA Isoforms by Exploiting Ionic Strength Effects during Ultrafiltration. *Biotechnol. Bioeng.* **2016**, *113*, 783–789. [CrossRef]
47. VIROVEK | BAC-TO-AAV Technology For Large Scale AAV Production. Available online: https://www.virovek.com (accessed on 29 August 2024).
48. Wright, J.F.; Le, T.; Prado, J.; Bahr-Davidson, J.; Smith, P.H.; Zhen, Z.; Sommer, J.M.; Pierce, G.F.; Qu, G. Identification of Factors That Contribute to Recombinant AAV2 Particle Aggregation and Methods to Prevent Its Occurrence during Vector Purification and Formulation. *Mol. Ther.* **2005**, *12*, 171–178. [CrossRef]
49. Evans, R.K.; Nawrocki, D.K.; Isopi, L.A.; Williams, D.M.; Casimiro, D.R.; Chin, S.; Chen, M.; Zhu, D.-M.; Shiver, J.W.; Volkin, D.B. Development of Stable Liquid Formulations for Adenovirus-Based Vaccines. *J. Pharm. Sci.* **2004**, *93*, 2458–2475. [CrossRef]
50. Qu, W.; Wang, M.; Wu, Y.; Xu, R. Scalable Downstream Strategies for Purification of Recombinant Adeno-Associated Virus Vectors in Light of the Properties. *Curr. Pharm. Biotechnol.* **2015**, *16*, 684–695. [CrossRef]
51. Carpenter, A.E.; Jones, T.R.; Lamprecht, M.R.; Clarke, C.; Kang, I.H.; Friman, O.; Guertin, D.A.; Chang, J.H.; Lindquist, R.A.; Moffat, J.; et al. CellProfiler: Image Analysis Software for Identifying and Quantifying Cell Phenotypes. *Genome Biol.* **2006**, *7*, R100. [CrossRef]
52. Stirling, D.R.; Swain-Bowden, M.J.; Lucas, A.M.; Carpenter, A.E.; Cimini, B.A.; Goodman, A. CellProfiler 4: Improvements in Speed, Utility and Usability. *BMC Bioinform.* **2021**, *22*, 433. [CrossRef]
53. Urabe, M.; Ding, C.; Kotin, R.M. Insect Cells as a Factory to Produce Adeno-Associated Virus Type 2 Vectors. *Human. Gene Ther.* **2002**, *13*, 1935–1943. [CrossRef] [PubMed]
54. Ebberink, E.H.T.M.; Ruisinger, A.; Nuebel, M.; Thomann, M.; Heck, A.J.R. Assessing Production Variability in Empty and Filled Adeno-Associated Viruses by Single Molecule Mass Analyses. *Mol. Ther. Methods Clin. Dev.* **2022**, *27*, 491–501. [CrossRef] [PubMed]
55. Wada, M.; Uchida, N.; Posadas-Herrera, G.; Hayashita-Kinoh, H.; Tsunekawa, Y.; Hirai, Y.; Okada, T. Large-Scale Purification of Functional AAV Particles Packaging the Full Genome Using Short-Term Ultracentrifugation with a Zonal Rotor. *Gene Ther.* **2023**, *30*, 641–648. [CrossRef] [PubMed]
56. Chen, H.; Marino, S.; Ho, C.Y. 97. Large Scale Purification of AAV with Continuous Flow Ultracentrifugation. *Mol. Ther.* **2016**, *24*, S42. [CrossRef]
57. Lam, A.K.; Mulcrone, P.L.; Frabutt, D.; Zhang, J.; Chrzanowski, M.; Arisa, S.; Munoz, M.; Li, X.; Biswas, M.; Markusic, D.; et al. Comprehensive Comparison of AAV Purification Methods: Iodixanol Gradient Centrifugation vs. Immuno-Affinity Chromatography. *Adv. Cell Gene Ther.* **2023**, *2023*, 2339702. [CrossRef]
58. Imiołek, M.; Fekete, S.; Kizekai, L.; Addepalli, B.; Lauber, M. Fast and Efficient Size Exclusion Chromatography of Adeno Associated Viral Vectors with 2.5 Micrometer Particle Low Adsorption Columns. *J. Chromatogr. A* **2024**, *1714*, 464587. [CrossRef]
59. Fernandes, P.; Peixoto, C.; Santiago, V.M.; Kremer, E.J.; Coroadinha, A.S.; Alves, P.M. Bioprocess Development for Canine Adenovirus Type 2 Vectors. *Gene Ther.* **2013**, *20*, 353–360. [CrossRef] [PubMed]
60. Wagner, C.; Innthaler, B.; Lemmerer, M.; Pletzenauer, R.; Birner-Gruenberger, R. Biophysical Characterization of Adeno-Associated Virus Vectors Using Ion-Exchange Chromatography Coupled to Light Scattering Detectors. *Int. J. Mol. Sci.* **2022**, *23*, 12715. [CrossRef]
61. McIntosh, N.L.; Berguig, G.Y.; Karim, O.A.; Cortesio, C.L.; De Angelis, R.; Khan, A.A.; Gold, D.; Maga, J.A.; Bhat, V.S. Comprehensive Characterization and Quantification of Adeno Associated Vectors by Size Exclusion Chromatography and Multi Angle Light Scattering. *Sci. Rep.* **2021**, *11*, 3012. [CrossRef]

membranes MDPI

Article

Comparative Analysis of the Impact of Protein on Virus Retention for Different Virus Removal Filters

Mohammad A. Afzal, Joshua Peles and Andrew L. Zydney *

Department of Chemical Engineering, Pennsylvania State University, University Park 16802, PA, USA; maa6831@psu.edu (M.A.A.); jmp6368@psu.edu (J.P.)
* Correspondence: zydney@engr.psu.edu

Abstract: The performance of virus filters is often determined by the extent of protein fouling, which can affect both filtrate flux and virus retention. However, the mechanisms governing changes in virus retention in the presence of proteins are still not well understood. The objective of this work was to examine the effect of proteins on virus retention by both asymmetric (Viresolve® NFP and Viresolve® Pro) and relatively homogeneous (Ultipor® DV20 and Pegasus™ SV4) virus filtration membranes. Experiments were performed with bacteriophage ϕX174 as a model parvovirus and human serum immunoglobulin G (hIgG) as a model protein. The virus retention in 1 g/L hIgG solutions was consistently less than that in a protein-free buffer solution by between 1 to 3 logs for the different virus filters. The virus retention profiles for the two homogeneous membranes were very similar, with the virus retention being highly correlated with the extent of flux decline. Membranes prefouled with hIgG and then challenged with phages also showed much lower virus retention, demonstrating the importance of membrane fouling; the one exception was the Viresolve® Pro membrane, which showed a similar virus retention for the prefouled and pristine membranes. Experiments in which the protein was filtered after the virus challenge demonstrated that hIgG can displace previously captured viruses from within a filter. The magnitude of these effects significantly varied for the different virus filters, likely due to differences in membrane morphology, pore size distribution, and chemistry, providing important insights into the development/application of virus filtration in bioprocessing.

Keywords: virus filtration; protein fouling; monoclonal antibody; virus clearance

1. Introduction

Mammalian cell-based expression systems are increasingly used for the production of recombinant protein products. For example, Chinese hamster ovary cells are the most commonly employed cell line for the manufacture of monoclonal antibodies [1]. However, these mammalian cell systems are susceptible to virus contamination, including by the parvovirus minute virus of mice (MVM) [2]. The risks of virus contamination can be dramatically reduced by utilizing low-risk starting materials and testing in-process streams, but there is still a need to incorporate effective methods of virus inactivation and/or removal as part of the downstream purification process [3–5]. Virus filtration is widely accepted across the industry as a robust, size-based method for the removal of virus particles [6].

A number of studies have demonstrated that the virus removal capabilities of commercial virus filters can be significantly altered by the presence of a therapeutic protein. Regulatory agencies thus require that virus filter challenges be performed in the actual process intermediates [7]. For instance, Stuckey et al. [8] evaluated virus retention for the Planova™ 20N virus removal filter using 11 different process streams, with the log reduction value (LRV) ranging 2.86 to 7.15. There was no apparent correlation with the pH, product concentration, solution conductivity, physical characteristics of the product, or extent of fouling (as measured by the flux decay). In contrast, corresponding challenge experiments performed with the XMuLV retrovirus in the same process streams showed undetectable levels of virus in all permeate samples (LRV $\geq$ 6). Bolton et al. [9] noted a significant reduction in virus retention with process throughput for the Viresolve® NFP membrane, with a much greater decline in LRV for virus challenge experiments performed in the presence of proteins. The measured LRV was highly correlated with the extent of flux decline, suggesting that the primary effect of the protein was due to protein fouling, which was thought to cause a shift in flow due to the preferential blockage of the smaller pores within the filter.

Suh et al. [10] examined the retention of the MS2 bacteriophage by seven commercial virus filters using bovine serum albumin (BSA) as a model protein. The Viresolve® NFP membrane showed the greatest flux decline and the greatest reduction in LRV, with the virus retention decreasing to LRV $\approx$ 1 after only 70 L/m^2. The Pegasus™ SV4 membrane also showed a significant reduction in LRV, even though the flux decline for this filter was minimal. Virus breakthrough was not seen when the filters were prefouled with BSA, suggesting that the effects of proteins on virus retention are not simply due to flux decline. Subsequent work [11] showed that the Viresolve® Pro membrane provided the robust retention of the PP7 bacteriophage even under highly fouling conditions. Lute et al. [12] identified several instances of virus breakthrough for certain virus filters even in the absence of any significant flux decay, indicating that protein fouling is not a universally reliable predictor of viral clearance. Hongo-Hirasaki et al. [13] reported that PPV retention through the Planova™ 20N was unaffected by IgG concentration. Likewise, Leisi et al. [14] observed that MVM retention by the Planova™ 20N membrane was largely unaffected by protein fouling, but there was a significant reduction in virus retention for the Pegasus™ SV4 membrane when operated under highly fouling conditions. Afzal and Zydney [15] also found a large reduction in virus retention by the Pegasus™ SV4 membrane when challenged with solutions of human serum immunoglobulin G (hIgG) spiked with the bacteriophage ϕX174, with the data suggesting that the hIgG displaced previously captured ϕX174 from within the pores of this membrane. Kosiol et al. [16] demonstrated that hIgG could cause virus aggregation, particularly at low pH, leading to an increase in virus retention under these conditions. However, the retention of the PP7 bacteriophage at pH 9 was reduced in the presence of hIgG, which the authors attributed to competitive adsorption effects, with hIgG occupying binding sites that would otherwise capture PP7.

Further support for the presence of interactions between viruses, proteins, and membranes is provided by studies showing a shift in the location of virus capture in the presence of proteins. For example, Adan-Kubo et al. [17] showed that fluorescently labeled B19V were captured closer to the inner surface of the Planova™ 20N membrane in the presence of proteins, with the magnitude of this shift being greatest for experiments performed with albumin relative to antithrombin, haptoglobin, or hIgG. Similar results were obtained by Leisi et al. [14] for the Planova™ 20N and BioEX membranes using fluorescently labeled MVM in the presence of hIgG, but the MVM were captured farther into the depth of the Pegasus™ SV4 membrane. The authors hypothesized that this unusual behavior for the Pegasus™ SV4 membrane was due to competition between the protein and virus for capture sites within the filter.

Additional insights into the impact of protein fouling on virus retention have been obtained by "prefouling" membranes through the filtration of a protein solution before challenging the membranes with a protein-free solution of the virus. Afzal and Zydney [15] demonstrated that prefouling the Pegasus™ SV4 membrane with hIgG led to a 1-log reduction in virus capture that was nearly independent of the initial degree of fouling. Khan et al. [18] observed no change in virus retention when prefouling the Viresolve® NFP membrane with BSA to 30% flux decay, but they found an increase in virus retention for membranes fouled to 60% flux decay. Jackson et al. [19] found no impact on virus retention after prefouling the Ultipor® DV20 membrane with hIgG up to a 50% flux decay, which was similar to results obtained by Suh et al. [10] for the Viresolve® Pro and Pegasus™ SV4 membranes prefouled with BSA. The origin of these differences in performance is unclear.

The objective of this study was to obtain more detailed insights into the impact of proteins on virus retention through four different commercial virus filtration membranes: the asymmetric Viresolve® NFP and Viresolve® Pro membranes and the relatively homogeneous Ultipor® DV20 and Pegasus™ SV4 membranes. Data were obtained for filters challenged with a model virus both in the presence and absence of hIgG and after prefouling the membranes with virus-free protein solutions. Experiments were also performed in staged filtration experiments to identify the effect of proteins on previously captured viruses. These results provide new insights into how proteins can alter the virus retention characteristics of different virus removal filters.

2. Materials and Methods

2.1. Virus Filters

Virus challenges were performed with a single layer of four commercially available virus filtration membranes: the highly asymmetric Viresolve® NFP and Viresolve® Pro membranes from MilliporeSigma Corporation (Bedford, MA, USA) and the relatively homogenous Ultipor® DV20 and Pegasus™ SV4 membranes from Cytiva (Marlborough, MA, USA), as summarized in Table 1. Single layers of membrane were used in all experiments to make it easier to accurately measure the virus retention without having to use very highly concentrated suspensions of bacteriophages.

Table 1. Summary of the virus removal filters used in this work.

Filter	Manufacturer	Polymer	Morphology	Permeability
Viresolve® Pro	MilliporeSigma	Polyether-sulfone	Asymmetric	27–29 LMH/psi
Viresolve® NFP	MilliporeSigma	Polyvinylidene fluoride	Asymmetric	26–32 LMH/psi
Ultipor® DV20	Cytiva (formerly Pall Life Sciences)	Polyvinylidene fluoride	Relatively homogeneous	2.0–2.7 LMH/psi
Pegasus™ SV4	Cytiva (formerly Pall Life Sciences)	Polyvinylidene fluoride	Relatively homogeneous	2.8–3.7 LMH/psi

The Viresolve® NFP and Viresolve® Pro membranes were obtained as single-layer sheets directly from the manufacturer. The other two membranes were provided as 47 mm dual-layer discs, which were carefully separated into individual layers by immersing them in deionized water. Then, 25-mm discs were cut and placed in a polypropylene filter holder (Advantec MFS Inc., Dublin, CA, USA), providing a filtration area of 3.5 cm^2, with the skin/shiny side facing away from the feed (the orientation used in the commercial virus filtration devices). Limited experiments were conducted using the Viresolve® NFP membrane in an Amicon® Model 8010 ultrafiltration cell and using the Viresolve® Pro membrane in a custom Micro 40 device provided by MilliporeSigma (Bedford, MA, USA), both with a single layer of the membrane.

Filters were fouled with solutions of human serum immunoglobulin G (hIgG), obtained from Nova Biologics (Oceanside, CA, USA) as a lyophilized powder that was then mixed with 0.10 M phosphate buffered saline (PBS) to obtain the desired hIgG concentration. The hIgG was filtered first through 0.2 μm and then through 0.1 μm pore size filters (Cytiva, Marlborough, MA, USA) to remove any large protein aggregates prior to use.

2.2. Bacteriophage

Bacteriophage ϕX174 (ATCC-13706-B1, Manassas, VA, USA), which has been extensively used as a surrogate for mammalian viruses, was propagated in host *E. coli* (ATCC13706, Manassas, VA, USA) by mixing 100 μL of phage with *E. coli* grown in nutrient broth (NB) media prepared by mixing 8 g/L of nutrient broth powder (BD-234000) and

5 g/L of NaCl (Promega Corp., Madison, WI, USA) with deionized distilled water. The phages were propagated at 37 °C for ~6 h with gentle agitation in an incubator shaker (New Brunswick Scientific, Edison, NJ, USA). The phage suspension was centrifuged at 3500 rpm for 10 min (Beckman Coulter, Brea, CA, USA), with the phage collected in the supernatant and then filtered through a 0.2 μm sterile filter.

ϕX174 concentrations were determined using a plaque-forming assay by mixing 100 μL of the phage samples with 200 μL of *E. coli* and 3 mL of soft agar, which were then poured onto hard agar plates that were incubated at 37 °C for approximately 6 h in an inverted position. The number of plaques was determined by visual counting, with samples diluted to obtain a countable number of plaques on each plate (typically from 5 to 200). Additional details on the handling and analysis of the ϕX174 are provided by Afzal and Zydney [20].

2.3. *Virus Filtration*

Virus filtration was performed at both a constant filtrate flux, which was obtained by connecting a peristaltic pump (Masterflex, Vernon Hills, IL, USA) between the feed solution reservoir and the filter holder, and a constant transmembrane pressure, which was provided by connecting the filter holder to an air-pressurized feed reservoir (SR-TEK Ltd., Milton Keynes, UK). Limited experiments were conducted with the virus filter directly fed by a syringe pump (KD Scientific, Holliston, MA, USA) to maintain a constant flux.

The membranes were initially flushed with around 70 L/m^2 of 100 mM PBS at pH 7.4 to thoroughly wet the pore structure. The membrane's permeability was then evaluated from the slope of experimental data for the filtrate flux as a function of transmembrane pressure (TMP), using data at a minimum of three TMP, with the range of values for the different membrane samples provided in Table 1.

The feed reservoir was then filled with the challenge solution: a suspension of bacteriophage ϕX174, an hIgG solution, or an hIgG solution spiked with ϕX174. The virus retention data were typically reported in terms of the log reduction value (*LRV*):

$$LRV = -log_{10}\left(\frac{C_{filtrate}}{C_{Feed}}\right) \quad (1)$$

where $C_{filtrate}$ and C_{Feed} are the ϕX174 concentrations in grab samples of the filtrate and feed solutions, respectively.

3. Results and Discussion

Figure 1 shows the data obtained during constant pressure filtration at 210 kPa (30 psi) through the four virus removal filters for the ϕX174 challenges in PBS and 1 g/L hIgG solutions, all at pH 7.4 with 150 mM ionic strength. The ϕX174 feed concentration was 10^6 pfu/mL for the Viresolve® NFP, Ultipor® DV20, and Pegasus™ SV4 membranes, and it was 10^8 pfu/mL for the Viresolve® Pro membrane; the higher phage concentration used for the Viresolve® Pro membrane was required to obtain detectable phage concentrations in the permeate samples obtained with this highly retentive virus filter. Limited data obtained over a range of phage concentrations indicated that the measured virus retention was independent of the phage concentration, in good agreement with previous results for the Ultipor® DV20 membrane [19]. Repeat experiments showed good reproducibility, with LRV values within ±0.5 and flux decline within ±10% (see Figure S1).

The filtrate flux remained nearly constant for the experiments using ϕX174 in PBS with a less than 10% decline in flux (except for the Viresolve® NFP membrane, which showed closer to a 20% flux decline). In contrast, there were significant flux declines for the experiments with ϕX174 in the presence of hIgG, with a more than a 90% flux decline for the two asymmetric membranes and a more than 50% flux decline for the two homogeneous membranes. This behavior is discussed in more detail subsequently.

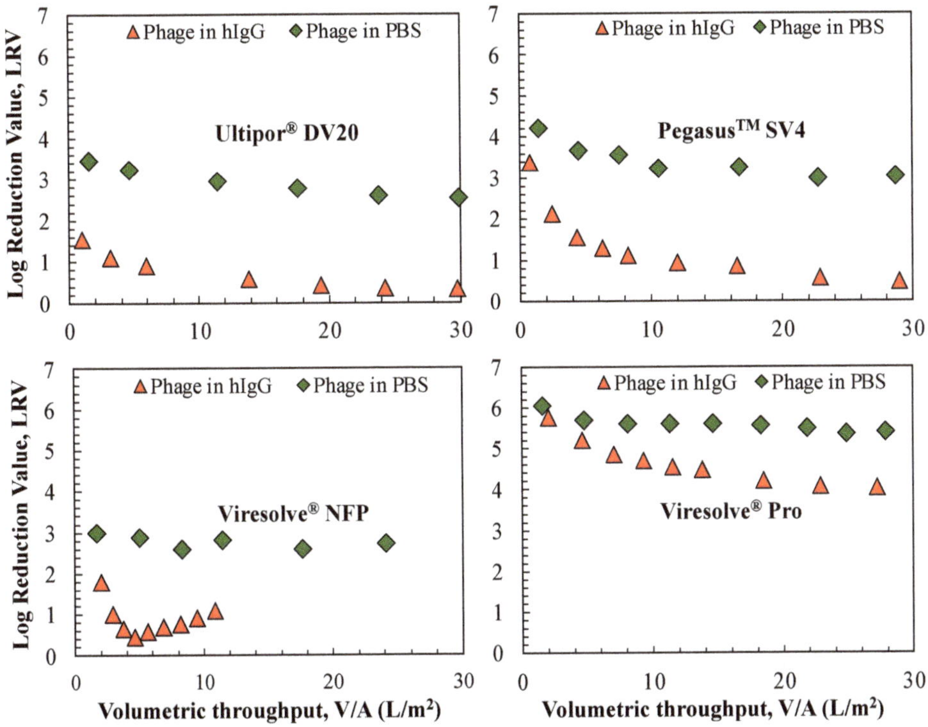

Figure 1. Virus retention during constant pressure filtration at 210 kPa for single layers of the Ultipor® DV20, Pegasus™ SV4, Viresolve® NFP, and Viresolve® Pro membranes for ϕX174 challenges using PBS or 1 g/L hIgG solutions. Data from single experiments; replicates are shown in Figure S1.

The measured virus retention in grab samples obtained over the course of the constant pressure filtration is shown in Figure 1 as a function of the volumetric throughput, equal to the cumulative filtrate volume divided by the filter area. The virus retention for the commercial virus filters would all be significantly higher than the LRV shown in Figure 1 since the Ultipor® DV20, Pegasus™ SV4, and Viresolve® Pro membranes are used as two-layers membranes in series and the Viresolve® NFP membrane is sold in a three-layer configuration. The decrease in LRV for the ϕX174 in PBS has previously been attributed to the accumulation of mobile viruses within the filter, a phenomenon referred to as internal polarization [15,19]. In all cases, the LRV obtained in the presence of hIgG was lower than that obtained in PBS, similar to results reported previously for the Viresolve® NFP [9] and Pegasus™ SV4 membranes [15]. This effect was relatively small for the Viresolve® Pro membrane, with the LRV in the presence of hIgG being about 1.3 logs smaller than that in PBS, but it was much more pronounced for the relatively homogeneous membranes; the LRV in the presence of hIgG approached zero for both the Ultipor® DV20 and Pegasus™ SV4 membranes, which is nearly 3 logs smaller than the values obtained in PBS. The ϕX174 retention for the Viresolve® NFP membrane showed an unusual behavior, with the LRV rapidly dropping to a value of 0.45 after the filtration of only 5 L/m^2 before increasing by about 0.5 logs over the subsequent 5 L/m^2. This may reflect the development of a filter cake or extensive pore blockage at very high degrees of fouling, which would cause an increase in ϕX174 retention during the latter stages of filtration. Khan et al. [18] presented a similar hypothesis for the increase in virus retention observed with viral stock solutions containing high levels of impurities.

Bolton et al. [9] previously hypothesized that the decline in LRV observed for the Viresolve® NFP membrane in the presence of proteins was directly related to the magnitude of the flux decline caused by protein fouling. This was examined for the four virus filters by replotting the data in Figure 1 as an explicit function of 1-J/J_o, where the values of the initial filtrate flux for each membrane (J_o) were evaluated based on data for the buffer flux at the same TMP. The results are shown in Figure 2; data from replicate experiments are plotted in Figure S2. The data for the two relatively homogeneous membranes, the Ultipor® DV20 and Pegasus™ SV4 membranes, tend to collapse to nearly a single curve when plotted in this fashion, although the virus retention decreased to nearly zero after only a 40% flux decline. This decline in LRV was likely related to the accumulation of ϕX174 in the capture zone of these homogeneous membranes in combination with the elimination of capture sites within the filter by the hIgG, either due to protein adsorption and/or the blockage of the size-restrictive pores. In contrast, the two highly asymmetric membranes, the Viresolve® Pro and Viresolve® NFP membranes, both showed very high degrees of fouling, but the virus retention for the Viresolve® Pro membrane remained at LRV > 4 while that for the Viresolve® NFP membrane sharply decreased, with the latter in good agreement with results obtained by Bolton et al. [9]. Note that the first data point for the Viresolve® NFP membrane is plotted at a flux decay of 88%, reflecting the very high degree of fouling that was obtained within the first grab sample (0.6 mL). The very high degree of fouling for the Viresolve® Pro and Viresolve® NFP membranes was a direct result of the highly asymmetric pore structure in these membranes, with all the protein fouling localized to the region with the highly restricted pores near the filter exit [21]. Note that the minimum in the LRV for the Viresolve® NFP membrane near the very end of the experiment occurred at a flux decline of 99.3%, which is a much higher degree of fouling than is typically studied in virus filter testing.

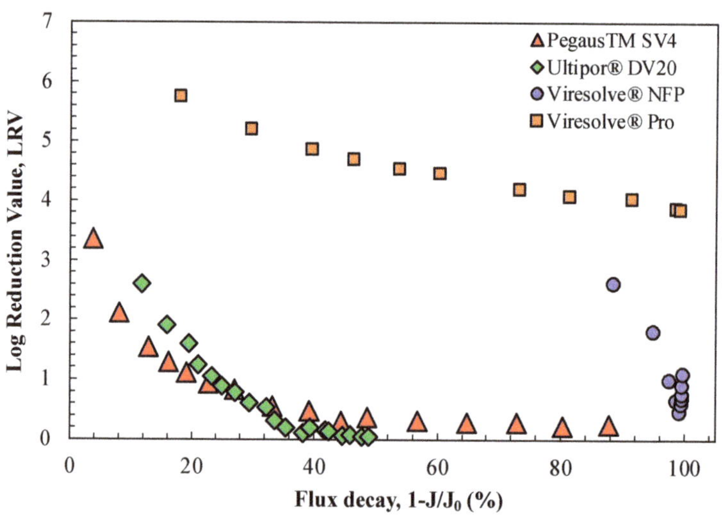

Figure 2. LRV as a function of flux decline for the experimental data from Figure 1.

Although most previous studies have attributed reductions in virus retention in the presence of proteins simply to the effects of protein fouling (and the resulting change in filtrate flux and pore size distribution), Afzal and Zydney [15] recently showed that hIgG could directly affect virus capture/retention by the Pegasus™ SV4 membrane. This was experimentally examined by performing virus challenge experiments at a constant filtrate flux (to eliminate the effect of flux decline on virus retention) for (a) clean membranes using ϕX174 in PBS, (b) membranes prefouled with hIgG and then challenged with ϕX174 in PBS, and (c) membranes challenged with a mixture of ϕX174 and hIgG. Given the very different permeabilities and fouling characteristics for the virus filters, these experiments were performed under slightly different conditions: the Ultipor® DV20 and Pegasus™ SV4 membranes were both operated at a flux of 20 LMH using 3 and 5 g/L hIgG solutions, respectively, the Viresolve® Pro membrane was run at 100 LMH with 0.25 g/L of hIgG, and the Viresolve® NFP membrane was run at 250 LMH with 0.05 g/L of hIgG. The prefouling was performed until the transmembrane pressure increased by a factor of 4 (i.e., until $P/P_o = 4$) for the Viresolve® Pro membrane and by a factor of 2.5 for the Pegasus™ SV4, Ultipor® DV20, and Viresolve® NFP membranes. The lower degree of fouling used for the experiments with the Pegasus™ SV4 and Ultipor® DV20 membranes was required because of the low permeability (and thus high initial transmembrane pressure) of these membranes. The experiments with the Viresolve® Pro membrane were conducted with a syringe pump to ensure that the filtrate flux remained constant.

The measured values of ϕX174 in the grab samples obtained at a throughput of 20 L/m^2 are summarized in Figure 3. In all cases, the highest virus retention was obtained for the clean membranes, while the lowest virus retention was obtained for the experiments in which the ϕX174 challenge was performed in the presence of hIgG, even though the experiments performed using the hIgG began with membranes that were "unfouled" and the pressure at a throughput of 20 L/m^2 was always less than or equal to that for the corresponding prefouled membrane. The difference in LRV between that in the prefouled membranes and that evaluated in the presence of hIgG was particularly pronounced for the relatively homogeneous Pegasus™ SV4 and Ultipor® DV20 membranes, with a more than 0.7-log difference in LRV. These homogeneous membranes show virus capture in a relatively larger portion of the filter depth [14,19,22], with the data suggesting that hIgG may out-compete ϕX174 for these capture sites. In contrast, the LRVs for the highly asymmetric Viresolve® NFP and Viresolve® Pro membranes were similar for the prefouled membranes and that evaluated in the presence of hIgG, with the Viresolve® Pro membrane showing robust virus retention for all three conditions. The origins of these differences in behavior are discussed in more detail subsequently.

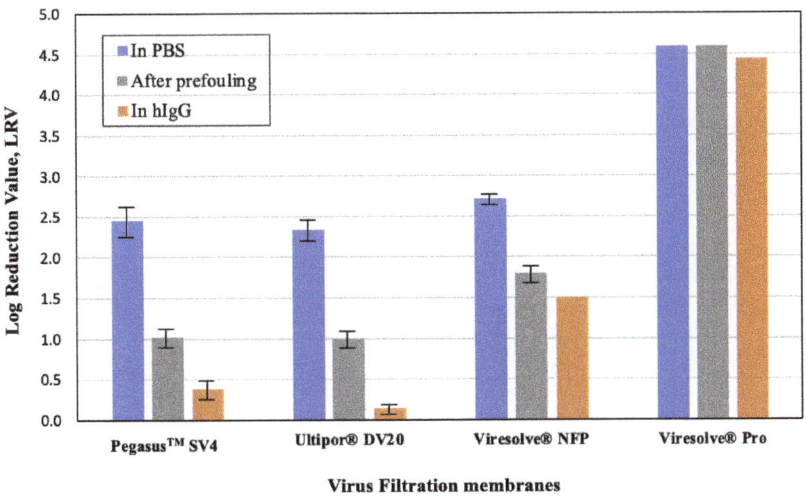

Figure 3. Virus retention for the different virus filtration membranes for single constant flux filtration experiments performed with ϕX174 in PBS through both clean and prefouled membranes and with ϕX174 in hIgG solutions. Prefouling was performed with the constant flux filtration of hIgG through the filters until $P/P_0 = 2.5$ for the Viresolve® NFP, Pegasus™ SV4, and Ultipor® DV20 membranes and to $P/P_0 = 4$ for the Viresolve® Pro membrane. Error bars represent the range of values for replicate experiments.

Additional insights were obtained by challenging the membranes with ϕX174 in PBS followed immediately by a virus-free 1 g/L hIgG solution, all at a constant pressure of 210 kPa. A trivalve was used to switch between the ϕX174 and hIgG to ensure that there was no disruption in the pressure/flow while switching between the feed solutions. The results from replicate experiments performed with each of the virus filters are plotted in Figure 4 as the normalized ϕX174 concentration, i.e., the ratio of the ϕX174 concentration in the permeate grab sample to that in the initial feed used for the phage challenge. The phage retention levels for the Ultipor® DV20, Pegasus™ SV4, and Viresolve® NFP membranes were nearly identical during the initial phage challenge, with the normalized ϕX174 concentration increasing by almost 1 log over the 35 L/m^2. In contrast, the Viresolve® Pro membrane had approximately 100-fold (2 logs) greater retention, although the normalized ϕX174 concentration also increased by nearly 1 log for this filter during the initial phage challenge.

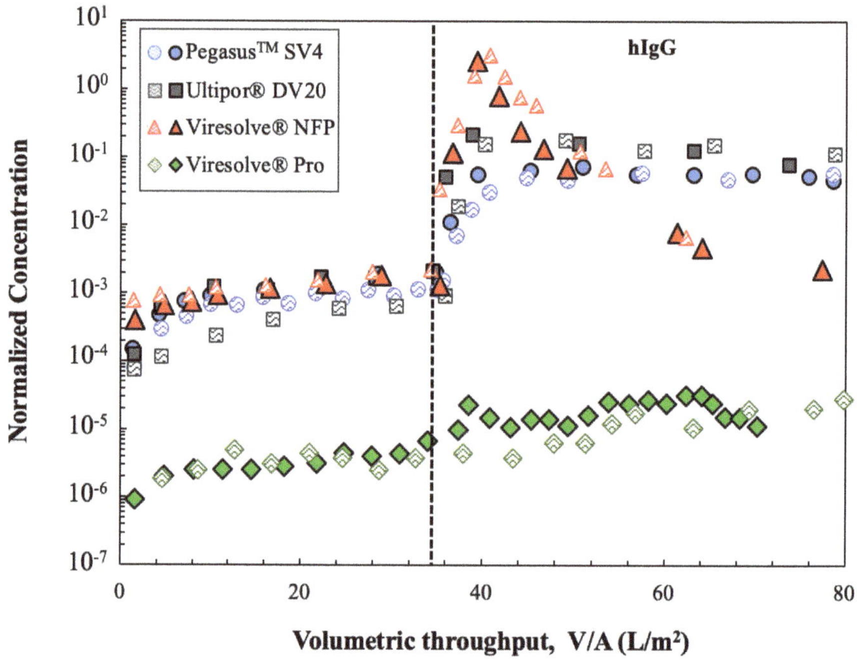

Figure 4. Normalized ϕX174 concentration (permeate concentration/initial feed concentration) as a function of volumetric throughput for replicate experiments involving different membranes for the filtration of ϕX174 followed by a 1 g/L hIgG solution at a constant pressure of 210 kPa. The feed was switched between ϕX174 and virus-free hIgG at 35 L/m^2 through a trivalve, as indicated by the dashed vertical line. Results from replicate experiments are shown by the shaded and filled symbols.

The introduction of hIgG caused a dramatic increase in the ϕX174 concentration in the permeate samples obtained through the Ultipor® DV20, Pegasus™ SV4, and Viresolve® NFP membranes, even though the hIgG solution contained no virus. Data obtained using the same experimental system but with PBS in the second half of the filtration showed no increase in ϕX174 concentration after switching from the phage suspension to pure buffer (shown subsequently in Figure 5). The increase in virus transmission upon switching to a hIgG solution was discussed by Afzal and Zydney [15] for the Pegasus™ SV4 membrane and was attributed to the displacement of previously captured phages by the hIgG. The ϕX174 concentration in the permeate samples for the Ultipor® DV20 and Pegasus™ SV4 membranes remained nearly constant during the hIgG challenge, with values that were approximately 100 times greater than that evaluated immediately before the introduction of the hIgG solution. The behavior of the Viresolve® NFP membrane was somewhat different, with an initial 1000-fold increase in the ϕX174 concentration followed by a very significant decline in permeate concentration over the course of the hIgG filtration. This reduction in phage concentration is likely related to the very rapid fouling of the Viresolve® NFP membrane by the hIgG solution, with the flux decreasing by more than 90% after only 3 L/m^2. In contrast, the ϕX174 retention by the Viresolve® Pro membrane was nearly unaffected by the switch to hIgG, with the ϕX174 concentration in the permeate remaining relatively stable throughout the hIgG filtration. It is important to note that the Ultipor® DV20, Pegasus™ SV4, and Viresolve® NFP membranes are all made of polyvinylidene fluoride (PVDF), while the Viresolve® Pro membrane is made of polyethersulfone, suggesting that the differences seen in Figures 3 and 4 may be related, at least in part, to the chemistry of the filters and the nature of the virus/protein/membrane interactions.

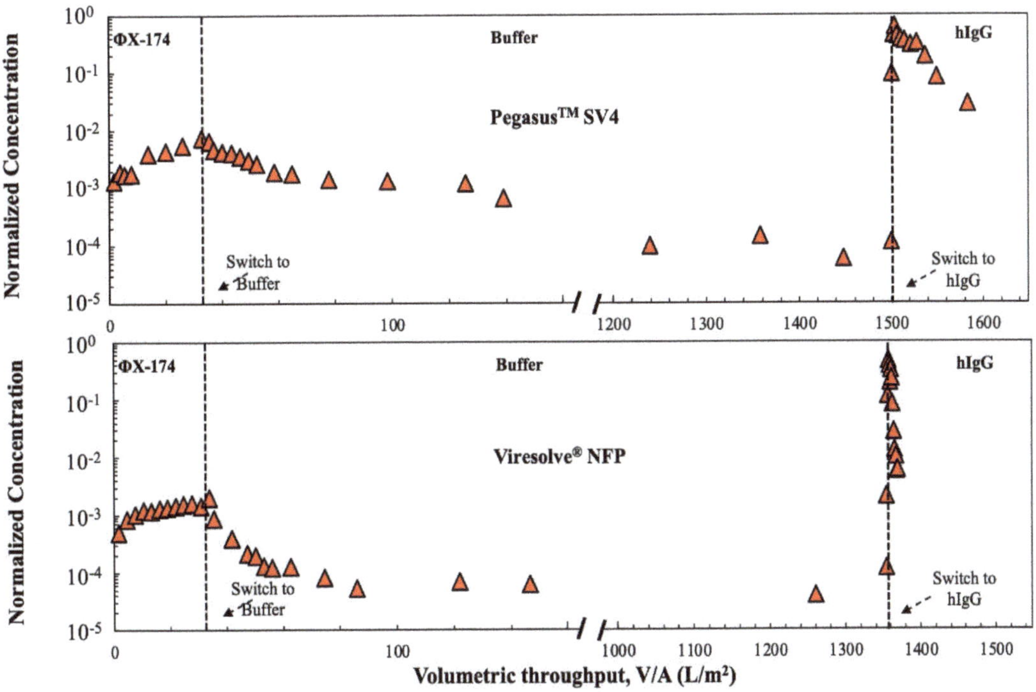

Figure 5. Normalized ϕX174 concentration in permeate samples collected through the Pegasus™ SV4 (**top panel**) and Viresolve® NFP (**bottom panel**) membranes during a 3-stage filtration process in which the filter was challenged first with ϕX174 in PBS followed by pure PBS (no phage) and then a virus-free hIgG solution, all at a constant transmembrane pressure of 210 kPa. The feed was switched between the different solutions through a trivalve, as indicated by the vertical dashed lines. Note the broken x-axis.

The different behavior of the highly asymmetric Viresolve® NFP and the relatively homogeneous Pegasus™ SV4 membranes were further examined using a three-stage filtration process in which the filters were first challenged with a suspension containing 10^6 pfu/mL ϕX174 in PBS for 35 L/m^2 throughput, followed immediately by an extended flush with just PBS and then a phage-free hIgG solution (1 g/L of hIgG for the Viresolve® NFP membrane and 5 g/L of hIgG for the Pegasus™ SV4 membrane). The switches in feed were again performed using a trivalve to eliminate any flow disruptions, with the pressure maintained at 210 kPa throughout the experiment. The results are shown in Figure 5. The extended buffer flush caused a gradual decline in the ϕX174 concentration for the Pegasus™ SV4 membrane, with the normalized phage concentration remaining around 10^{-4} over more than 1400 L/m^2 of filtration due to the very slow washout of previously retained phage during the buffer flush [15]. The buffer flush caused a more rapid initial decline in the ϕX174 concentration for the Viresolve® NFP membrane, with the normalized concentration remaining at about 4×10^{-5} over the remainder of the buffer flush. Although the extended buffer flush should have led to the removal of the large majority of the mobile ϕX174 that was retained within the membrane during the initial phage challenge [19], the switch to the hIgG solution still caused a rapid, nearly 4-log increase in the normalized ϕX174 concentration for both membranes despite the very different pore morphologies. This large spike in the ϕX174 transmission was likely due to the displacement of previously captured phages (that were not removed by the extended buffer flush) by the hIgG in the final stage of the filtration experiment, possibly reflecting the presence of competitive adsorptive interactions between the ϕX174/hIgG and the PVDF membranes, with adsorption of the hIgG leading to the desorption of the ϕX174.

The potential impacts of adsorptive interactions on ϕX174 retention were further examined by challenging the Viresolve® NFP and Pegasus™ SV4 membranes with ϕX174 at different solution pH values. Data were obtained for the Pegasus SV4 membrane in 10 mM PBS solutions to increase the effect of electrostatic interactions, with the pH adjusted by adding HCl or NaOH as needed. The results are summarized in Table 2. The initial LRV after 10 L/m^2 of throughput for the Viresolve® NFP membrane significantly increased at low pH levels, i.e., under conditions where the ϕX174 and membranes had opposite electrical charges. There was also a small increase in LRV with decreasing pH for the Pegasus™ SV4 membrane, although this effect was smaller than that seen with the Viresolve® NFP membrane.

Table 2. Effect of solution pH on virus retention by the Viresolve® NFP and Pegasus™ SV4 membranes.

pH	Viresolve® NFP LRV	Pegasus™ SV4 LRV
4.0–4.9	4.4 ± 0.2	3.1 ± 0.2
7.4	2.7 ± 0.2	2.8 ± 0.2
10	2.4 ± 0.2	2.4 ± 0.2

The effects of hIgG concentration on ϕX174 retention for the Viresolve® NFP and Pegasus™ SV4 membranes are examined in Figure 6. The experiments with the Viresolve® NFP membrane were performed at lower hIgG concentrations due to the very high degree of fouling observed with this filter. The ϕX174 retention by the Viresolve® NFP membrane decreased with increasing hIgG concentration, similar to the results presented by Bolton et al. [9], with the LRV decreasing to essentially zero for the experiments with 0.08 and 0.2 g/L hIgG solutions. Only the data at the highest hIgG concentration showed a minimum in LRV at an intermediate throughput. The data for the Pegasus™ SV4 membrane showed only a 2-log reduction in LRV when the hIgG concentration was 0.2 g/L, but the LRV decline was much more pronounced for the runs at hIgG concentrations of 1 g/L or higher. In addition, the data at 7.5 and 10 g/L hIgG showed minima in the phage retention, similar to the results for the Viresolve® NFP membrane at 1 g/L.

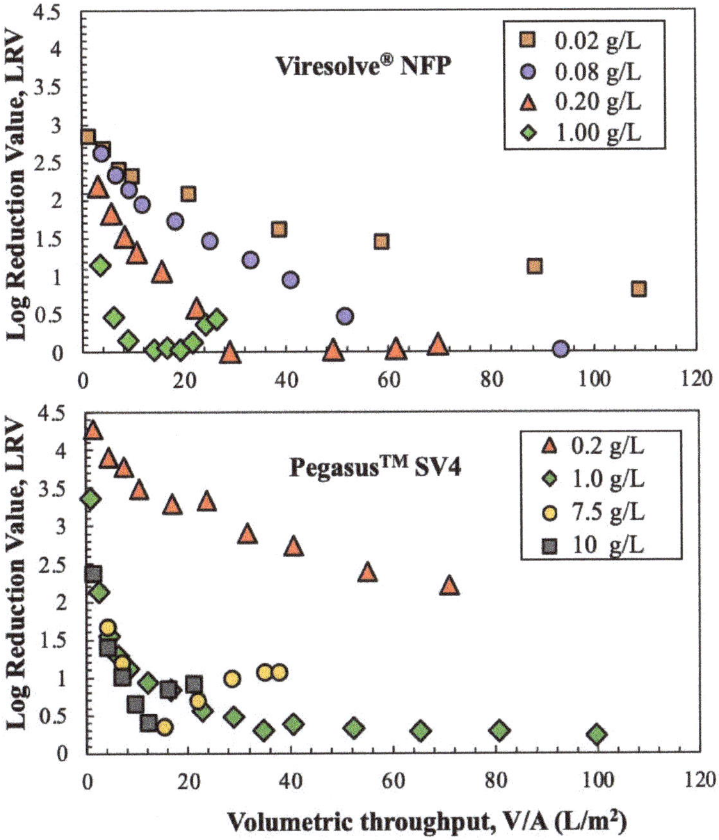

Figure 6. Virus retention as a function of hIgG concentration for protein solutions with 10^6 pfu/mL of phages. Data were obtained for single experiments at each hIgG concentration involving the Viresolve® NFP (**top panel**) and Pegasus™ SV4 (**bottom panel**) membranes, all at a constant pressure of 210 kPa.

4. Conclusions

The results presented in this study provide detailed insights into the virus retention performance of four different commercially available virus filtration membranes. Similar to previously reported studies, virus retention was reduced in the presence of proteins compared with that obtained in a protein-free buffer. The difference in LRV was nearly 3 logs for the relatively homogenous Ultipor® DV20 and Pegasus™ SV4 membranes, while the highly asymmetric Viresolve® Pro membrane had only about 1.3 logs lower retention (even for the heavily fouled membrane). The Viresolve® NFP membrane also had a lower LRV in the presence of hIgG, but in this case, the virus retention in the presence of hIgG showed a distinct minimum, with the increase in virus retention at a very high flux decline likely due to the blockage of the pore space within the depth of the Viresolve® NFP membrane (upstream of the size-selective skin layer) that provides additional resistance to transport. The LRV profiles for the Ultipor® DV20 and Pegasus™ SV4 membranes were well-correlated with the extent of flux decline, suggesting that this behavior is governed in large part by the relatively homogeneous structure of these two filters.

Experiments performed with prefouled membranes at constant flux showed that only part of the reduction in LRV was due to a change in the membrane pore size distribution due to protein fouling; data for the Viresolve® Pro membrane showed nearly identical virus retention levels for the pristine and prefouled membranes. The different behavior for the different virus filters is likely related to differences in the underlying pore structure/properties of the membranes, similar to the different effects of protein fouling on the location of virus capture as examined by confocal microscopy [14] and the visualization of gold nanoparticles by SEM [21]. Note that the Viresolve® Pro membrane is made of polyethersulfone, while the other three virus filters examined in this work were all made of PVDF, suggesting that the more robust virus retention for the Viresolve® Pro membrane may be directly related to its surface chemistry.

Sequential filtration experiments in which the filters were first challenged with a virus and then with hIgG (without virus) showed a sharp increase in virus concentration in permeate samples obtained immediately after switching to the protein. This increase in virus transmission was due to the displacement of viruses that were previously captured within the membrane by the protein, a phenomenon that was first identified by Afzal and Zydney [15] for the Pegasus™ SV4 membrane. This effect was most pronounced for the Viresolve® NFP membrane and least significant for the Viresolve® Pro membrane, suggesting that this behavior is not related to the asymmetric (versus homogeneous) structure of the membrane. This increase in virus transmission was also apparent even after extensively flushing the virus filter with buffer (using more than 1000 L/m^2 of buffer). Thus, it is critically important that any pre-flushing of virus filters be performed with completely virus-free buffer solutions to ensure that the product is not contaminated with viruses that were captured during the filter preparation.

The large reduction in LRV for the Ultipor® DV20, Pegasus™ SV4, and Viresolve® NFP membranes in the presence of proteins raises concerns about the effectiveness of these filters in providing the high level of virus removal required in bioprocessing. Similar effects have been previously reported. For example, Lute et al. [12] reported an LRV ≤ 0.5 for the Ultipor® DV20 membrane using both ϕX174 and PP7 in the presence of a 2.5 mg/mL solution of a therapeutic protein, while Bolton et al. [9] found an LRV ≤ 1 for the Viresolve® NFP using ϕX174 when measured in a 1 mg/mL solution of hIgG. However, it is important to note that the data obtained in this study were for single layers of membrane, while the commercially available flat-sheet virus filters typically use two or three layers of virus-retentive membranes. Peles et al. [23] showed that the hIgG fouling of the Viresolve® Pro membrane occurs almost entirely in the upper layer of the two-layer filter, suggesting that virus retention by the lower layer(s) of membrane might be largely unaffected by protein fouling (but might still be affected by the displacement of viruses in the presence of proteins). In addition, the hIgG fouling of virus filters appears to be dominated by the presence of irreversible aggregates [23,24], while many monoclonal antibodies show flux decline due to the formation of reversible associations. Future studies will be required to understand the magnitude of the effect of proteins on virus retention by these multilayer filters using actual therapeutic proteins with different fouling characteristics to more fully understand how the properties of the proteins and the key morphological features of the virus filters, such as pore tortuosity, void volume fraction, pore size distribution, and pore interconnectivity, govern virus filter performance.

Supplementary Materials: The following supporting information can be downloaded at: https://www.mdpi.com/article/10.3390/membranes14070158/s1. Figure S1: Virus retention during constant pressure filtration at 210 kPa for single layers of the Ultipor® DV20, Pegasus™ SV4, Viresolve® NFP, and Viresolve® Pro membranes for replicate experiments involving ϕX174 challenges using PBS or 1 g/L hIgG solutions. Results from replicate experiments are represented as filled and shaded symbols. Figure S2: LRV as a function of flux decline for the replicate experiments involving phage in hIgG from Figure S1.

Author Contributions: Conceptualization, A.L.Z.; methodology, M.A.A., J.P. and A.L.Z.; data curation, M.A.A. and J.P.; writing—original draft preparation, M.A.A.; writing—review and editing, A.L.Z.; supervision, A.L.Z.; funding acquisition, A.L.Z. All authors have read and agreed to the published version of the manuscript.

Funding: This work was supported through the Membrane Applications Science and Technology (MAST) Center, which is funded by grant numbers 1841474 and 2310832 from the NSF IUCRC program.

Institutional Review Board Statement: Not applicable.

Data Availability Statement: The data presented in this study are available on request from the corresponding author.

Acknowledgments: The authors would like to thank Brian Chettle for assisting with some of the experimental studies and MilliporeSigma for the donation of the Viresolve® NFP and Viresolve® Pro membranes.

Conflicts of Interest: The authors declare no conflict of interest.

References

1. Walsh, G. Biopharmaceutical Benchmarks 2014. *Nat. Biotechnol.* **2014**, *32*, 992–1000. [CrossRef] [PubMed]
2. Merten, O.W. Virus contaminations of cell cultures—A biotechnological view. *Cytotechnology* **2002**, *39*, 91–116. [CrossRef] [PubMed]
3. International Conference on Harmonisation, Guidance on Viral Safety Evaluation of Biotechnology Products Derived from Cell Lines of Human or Animal Origin, Availability—FDA. *Federal Register Notices.* **1998**, *63*, 51074–51084. Available online: https://www.federalregister.gov/documents/1998/09/24/98-25569/international-conference-on-harmonisation-guidance-on-viral-safety-evaluation-of-biotechnology (accessed on 14 July 2024).
4. Shukla, A.A.; Aranha, H. Viral Clearance for Biopharmaceutical Downstream Processes. *Pharm. Bioprocess.* **2015**, *3*, 127–138. [CrossRef]
5. Barone, P.W.; Wiebe, M.E.; Leung, J.C.; Hussein, I.T.M.; Keumurian, F.J.; Bouressa, J.; Brussel, A.; Chen, D.; Chong, M.; Dehghani, H.; et al. Viral Contamination in Biologic Manufacture and Implications for Emerging Therapies. *Nat. Biotechnol.* **2020**, *38*, 563–572. [CrossRef] [PubMed]
6. Johnson, S.A.; Chen, S.; Bolton, G.; Chen, Q.; Lute, S.; Fisher, J.; Brorson, K. Virus Filtration: A Review of Current and Future Practices in Bioprocessing. *Biotechnol. Bioeng.* **2022**, *119*, 743–761. [CrossRef] [PubMed]
7. PDA. Virus filtration. Technical report No. 41. *PDA J. Pharm. Sci. Technol.* **2005**, *59*, 8–42.
8. Stuckey, J.; Strauss, D.; Venkiteshwaran, A.; Gao, J.; Luo, W.; Quertinmont, M.; O'Donnell, S.; Chen, D. A novel approach to achieving modular retrovirus clearance for a parvovirus filter. *Biotechnol. Prog.* **2013**, *30*, 79–85. [CrossRef]
9. Bolton, G.; Cabatingan, M.; Rubino, M.; Lute, S.; Brorson, K.; Bailey, M. Normal-flow Virus Filtration: Detection and Assessment of the Endpoint in Bioprocessing. *Biotechnol. Appl. Biochem.* **2005**, *42*, 133–142. [CrossRef]
10. Suh, D.; Jin, H.; Park, H.; Lee, C.; Cho, Y.H.; Baek, Y. Effect of Protein Fouling on Filtrate Flux and Virus Breakthrough Behaviors during Virus Filtration Process. *Biotechnol. Bioeng.* **2023**, *120*, 1891–1901. [CrossRef]
11. Suh, D.; Jin, H.; Lee, C.; Baek, Y. Virus breakthrough behavior during virus filtration: Effects of virus/protein concentration and operating conditions. *Biochem. Eng. J.* **2023**, *198*, 109031. [CrossRef]
12. Lute, S.; Bailey, M.; Combs, J.; Sukumar, M.; Brorson, K. Phage Passage after Extended Processing in Small-virus-retentive Filters. *Biotechnol. Appl. Biochem.* **2007**, *47*, 141–151. [CrossRef] [PubMed]
13. Hongo-Hirasaki, T.; Komuro, M.; Ide, S. Effect of antibody solution conditions on filter performance for virus removal filter planovaTM 20N. *Biotechnol. Prog.* **2010**, *26*, 1080–1087. [CrossRef]
14. Leisi, R.; Rostami, I.; Laughhunn, A.; Bieri, J.; Roth, N.J.; Widmer, E.; Ros, C. Visualizing Protein Fouling and Its Impact on Parvovirus Retention within Distinct Filter Membrane Morphologies. *J. Membr. Sci.* **2022**, *659*, 120791. [CrossRef]
15. Afzal, M.A.; Zydney, A.L. Impact of Proteins and Protein Fouling on Virus Retention during Virus Removal Filtration. *Biotechnol. Bioeng.* **2024**, *121*, 710–718. [CrossRef] [PubMed]
16. Kosiol, P.; Kahrs, C.; Thom, V.; Ulbricht, M.; Hansmann, B. Investigation of virus retention by size exclusion membranes under different flow regimes. *Biotechnol. Prog.* **2018**, *35*, e2747. [CrossRef]
17. Adan-Kubo, J.; Tsujikawa, M.; Takahashi, K.; Hongo-Hirasaki, T.; Sakai, K. Microscopic visualization of virus removal by dedicated filters used in biopharmaceutical processing: Impact of membrane structure and localization of captured virus particles. *Biotechnol. Prog.* **2019**, *35*, e2875. [CrossRef]
18. Khan, N.Z.; Parrella, J.J.; Genest, P.W.; Colman, M.S. Filter Preconditioning Enables Representative Scaled-down Modelling of Filter Capacity and Viral Clearance by Mitigating the Impact of Virus Spike Impurities. *Biotechnol. Appl. Biochem.* **2009**, *52*, 293–301. [CrossRef]
19. Jackson, N.B.; Bakhshayeshi, M.; Zydney, A.L.; Mehta, A.; Van Reis, R.; Kuriyel, R. Internal Virus Polarization Model for Virus Retention by the Ultipor®VF Grade DV20 Membrane. *Biotechnol. Prog.* **2014**, *30*, 856–863. [CrossRef]
20. Afzal, M.A.; Zydney, A.L. Effect of filtrate flux and process disruptions on virus retention by a relatively homogeneous virus removal membrane. *Biotechnol. Prog.* **2022**, *38*, e3255. [CrossRef]
21. Fallahianbijan, F.; Giglia, S.; Carbrello, C.; Bell, D.; Zydney, A.L. Impact of Protein Fouling on Nanoparticle Capture within the Viresolve® Pro and Viresolve® NFP Virus Removal Membranes. *Biotechnol. Bioeng.* **2019**, *116*, 2285–2291. [CrossRef] [PubMed]

22. Leisi, R.; Widmer, E.; Gooch, B.; Roth, N.J.; Ros, C. Mechanistic insights into flow-dependent virus retention in different nanofilter membranes. *J. Membr. Sci.* **2021**, *636*, 119548. [CrossRef]
23. Peles, J.; Cacace, B.; Carbrello, C.; Giglia, S.; Zydney, A.L. Protein Fouling during Constant-flux Virus Filtration: Mechanisms and Modeling. *Biotechnol. Bioeng.* **2023**, *120*, 3357–3367. [CrossRef] [PubMed]
24. Bieberbach, M.; Kosiol, P.; Seay, A.; Bennecke, M.; Hansmann, B.; Hepbildikler, S.; Thom, V. Investigation of Fouling Mechanisms of Virus Filters during the Filtration of Protein Solutions Using a High Throughput Filtration Screening Device. *Biotechnol. Prog.* **2019**, *35*, e2776. [CrossRef]

membranes

MDPI

Review

Toward Rational Design of Ion-Exchange Nanofiber Membranes: Meso-Scale Computational Approaches

Inci Boztepe [1], Shuaifei Zhao [1], Xing Yang [2] and Lingxue Kong [1,*]

[1] Institute for Frontier Materials, Deakin University, Waurn Ponds, Geelong, VIC 3216, Australia; iboztepe@deakin.edu.au (I.B.); s.zhao@deakin.edu.au (S.Z.)

[2] Department of Chemical Engineering, KU Leuven University, Celestijnenlaan 200F, B-3001 Leuven, Belgium; xing.yang@kuleuven.be

* Correspondence: lingxue.kong@deakin.edu.au; Tel.: +61-3-5227-2087

Abstract

This review highlights the growing relevance of ion-exchange nanofibrous membranes (IEX-NFMs) in membrane chromatography (MC) for protein purification, emphasising their structural advantages such as high porosity, tunable surface functionality, and low-pressure drops. While the adsorption of IEX-NFMs in MC is expanding due to their potential for high throughput and rapid mass transfer, a critical limitation remains: the precise binding capacity of these membranes is not well understood. Traditional experimental methods to evaluate protein–membrane interactions and optimise binding capacities are labour-intensive, time-consuming, and costly. Therefore, this review underscores the importance of computational modelling as a viable predictive approach to guide membrane design and performance prediction. Yet major obstacles persist, including the challenge of accurate representation of the complex and often irregular pore structures, as well as limited and/or oversimplified adsorption models. Along with molecular-scale simulations such as molecular dynamics (MD) simulations and quantum simulations, meso-scale simulations can provide insight into protein–fibre and protein–protein interactions under varying physicochemical conditions for larger time scales and lower computational burden. These tools can help identify key parameters such as binding accessibility, ionic strength effects, and surface charge density, which are essential for the rational design and performance prediction of IEX-NFMs. Moreover, integrating simulations with experimental validation can accelerate optimisation process while reducing cost. This technical review sets the foundation for a computationally driven design framework for multifunctional IEX-NFMs, supporting their use in next-generation chromatographic separations and broadening their applications in bioprocessing and analytical biotechnology.

Keywords: protein binding; ion-exchange nanofiber membrane; computational approaches; meso-scale simulations; CFD; DEM

1. Introduction

Membrane chromatography (MC) is an alternative process used in protein purification based on affinity, ion exchange, and hydrophobic interaction separation. In MC, stacks of microporous or macroporous functionalised polymeric membranes with larger pore sizes in a range from 0.65 to 3 μm [1] are placed in the capsule as an adsorptive medium [2]. While the stationary phase is the chromatographic medium or the adsorbent, the mobile phase is the solution that flows through the column [3,4]. MC can be operated in either batch or flow-through modes. Continuous or flow-through adsorption provides higher productivity and better quality compared to batch adsorption. The transition from batch to continuous processing is relatively straightforward, requiring only minimal modifications like adjustments in column size and flow rate [5]. Higher flow rates can be used due to the convective nature of the mass transfer through membranes, resulting in almost flow-independent binding.

In the ion-exchange process, separation is governed by electrostatic interactions between proteins and charged groups on the membrane, depending on the pI value of the protein [6]; cation-exchange membranes bind positively charged proteins, while anion-exchange membranes bind negatively charged proteins [7]. Hydrophobic interaction membrane chromatography relies on interactions between hydrophobic ligands (e.g., butyl, octyl, phenyl) and non-polar regions on protein surfaces, typically under high salt conditions to exclude polar impurities [8–10]. Affinity membrane chromatography utilises immobilised ligands that selectively recognise and bind target proteins via specific biorecognition mechanisms [7]. Mixed-mode membrane chromatography combines multiple interaction types such as ion-exchange, hydrophobic interactions, and hydrogen bonding. A combination of ion exchange and hydrophobic interactions can usually achieve high selectivity and sensitivity [9]. However, mixed-mode chromatography media preparation usually requires a complex process and makes it difficult to control ligand density [11]. In Table 1, the most common chromatography methods for protein separation can be found [3].

Table 1. Protein separation by various modes and their mechanisms.

Separation Type	Method
Ion exchange	Reversible adsorption of a charged protein to ion-exchange matrices containing oppositely charged covalently attached side groups [4]
Hydrophobic interaction	Adsorbents, which have covalently attached hydrophobic groups Hydrophilic region exposure is promoted by decreased ionic strength [12]
Affinity	Specific interaction between the immobilised ligand and the binding site on the target molecule [13] Selective and efficient target molecule capturing [14] Most selective chromatography types among others [12]
Mixed-mode	Selective protein adsorption by the combination of several separation methods and membrane adsorbents [8]

The MC process has several distinctive advantages over the conventional packed column, such as high flexibility, high flow rates, low-pressure drop, scalability, and potential for high-throughput operation [15,16]. Disposable MC capsules are especially useful in small- to preparatory-scale bioprocessing as they eliminate the need for column cleaning, packing, and process revalidation, making them suitable for small-scale production in early clinical studies [17–19]. Also, due to the high modularity of membrane-based systems, large volumetric capacities and high velocities can be employed in MC [20,21].

 https://doi.org/10.3390/membranes16010005

The mass transfer mechanism is the distinctive factor between column and membrane chromatography. The solute transport in membrane chromatography mainly occurs through convection to the membrane's internal surface, which is a much faster mechanism compared to the intraparticle diffusion of proteins through the pores in column chromatography [1,22]. This convective transport effect results in a much lower mass transfer resistance [22].

The main limitation to the large-scale industrial application of MC is its relatively low binding capacity [23]. This issue arises primarily from restricted access to functional sites, often caused by inadequate pore connectivity and flow maldistribution within the membrane matrix [18]. Membranes with larger pore sizes generally possess smaller volume-specific surface areas compared to those with smaller pores at the same porosity, which reduces available binding sites. Nevertheless, in some cases, non-uniform porosity, membrane thickness, and ligand grafting can lead to variable flow resistance within the porous media, which could result in early saturation of binding sites in large pores and a decrease in dynamic binding capacity. These properties should be treated carefully for the sake of maintaining the high-throughput membranes in protein purification.

To maintain high throughput in protein purification, careful control of these structural parameters is essential. Enhancing the binding capacity of adsorptive media, therefore, requires material-level innovations. Recent developments include the use of next-generation matrices such as hydrogels and nanofibres, which provide higher specific surface areas, increased ligand densities, and optimised 3D binding environments [19]. Advances in nanotechnology have further enabled the fabrication of nanofibrous membranes with interconnected pore networks and high porosity, significantly improving mass transfer and accessibility to active sites [18,19,24,25]. Owing to their exceptionally high specific surface area (~10–40 $m^2 \cdot g^{-1}$), ion-exchange nanofibre membranes represent a promising alternative for enhancing protein-binding performance in membrane chromatography [18].

Despite significant progress in ion-exchange nanofibre membranes for protein binding, the field still lacks a unified framework that links membrane structure, transport phenomena, and performance in a way that enables rational and predictive design. Several reviews have provided valuable insights into ion-exchange nanofibre membranes, including a comprehensive discussion on design principles, synthesis methods, and an application performance study of electrospun nanofibrous membranes [15], as well as fundamental aspects, scalable fabrication techniques, and broad application areas of IEX nanofibres [26]. More recently, reviews have provided innovative approaches for developing novel membranes with properties tailored to the future needs of downstream processing [27], detailed assessments of next-generation high-throughput ion-exchange membranes and the factors influencing their performance [1], and examined the rational selection of membrane materials and structural properties for efficient bind-elute antibody capture [28].

As the field moves toward next-generation chromatographic membranes, optimising nanofibrous pore structures and surface functionalities will be crucial to enhancing adsorption performance. A deeper understanding of how pore architecture, fibre morphology, and surface modifications affect mass transport is essential [15]. As experimentally evaluating all parameters that influence membrane performance is time-consuming and resource-intensive, computational simulations offer a powerful alternative through modelling flow, adsorption kinetics, and mass transport mechanisms and provide critical insights into guiding membrane design. These simulations are particularly valuable for understanding how pore structure and surface properties affect separation mechanisms in ion-exchange processes, directly linking membrane characteristics to protein-binding efficiency. Systematic studies of individual membrane characteristics and their impact on binding capacity remain limited in the literature. This review, therefore, aims to establish a

 https://doi.org/10.3390/membranes16010005

rational design framework in which computational approaches facilitate the development of next-generation ion-exchange membranes tailored for efficient protein-binding and chromatographic applications.

2. Protein Adsorption on Ion-Exchange (IEX) Nanofibre Membranes

Recently, modified polymeric membranes have gained prominence in downstream processing, particularly in biological filtration units, due to their enhanced performance in protein separation [29]. Among various fabrication methods, electrospinning has attracted significant attention in protein purification for its precise control over fibre assembly and morphology [15,30,31]. Electrospun nanofibrous membranes are especially promising in ion-exchange membrane chromatography (IEMC) because of their high specific surface area (SSA), highly tortuous porous structures, and robust mechanical stability, all of which promote efficient protein adsorption [32]. Additionally, uniform nanofibres can be produced in a continuous process with control over the structure [33]. Specifically, electrospun nanofibre membranes enable chemical/physical functionalization that can result in high separation efficiencies [34]. Because of functionalization, the majority of binding sites on the surfaces of these nanofibres are accessible and external, resulting in a high adsorption capacity. This leads to numerous accessible binding sites on the nanofibres, therefore increasing the binding capacity of membranes.

Protein adsorption on the IEX nanofibre membrane occurs through static or dynamic adsorption. In static adsorption, the membrane acts as an adsorbent, with surface area being the key determinant of adsorption capacity. In dynamic adsorption, the membrane integrates adsorption and filtration within a single-pass flow [26]. Reported dynamic binding capacities (DBCs) for commercially available micron-size fibrous membranes range from 29 to 70 mg/mL bovine serum albumin (BSA) for anion exchange membranes and 29 to 47 mg/mL Lysozyme for cation-exchange membranes [1]. In comparison, electrospun nanofibre membranes demonstrate notably higher binding capacities in membrane chromatography, as summarised in Table 2.

Table 2. Performance of nanofibrous membranes in MC for the purification of targeted proteins across various studies. "*" is for the unit of mg/mL. Unless it is marked "*", it is mg/g.

Ref.	Nanofibrous Membrane	Fibre Diameter (d_f) (nm)	Water Contact Angle (WCA) (°)	Mean Pore Size (μm)	Specific Surface Area (SSA) (m^2/g)	Target Protein	pH	Initial Concentration (mg/mL)	Binding Capacity (mg/g or mg/mL *)
[35]	PAN-pAQ	353 ± 69–673 ± 36	115 ± 5–132 ± 0.6	0.9 ± 0.1–1.7 ± 0.2	4.4–9.3	BSA	7.5	-	98 ± 8–166 ± 2 (SBC)
[36]	PSf-GMA-DEA PAN-GMA-DEA	2500 ± 460 (PSf) 150–340 (PAN)	128 ± 4.8 (PSf) 73.1 ± 3.5 (PAN)	-	-	BSA	7.0	0.5–3	201.3 ± 4.8 * (PSf) 87.2 ± 2.6 * (PAN) (DBC) 260 * (PSf) 100 * (PAN) (SBC)
[16]	EVOH-CCA	562	120	-	2.52	Lysozyme	4–10	1	250 (DBC)
[37]	SA-PEO	150	-	0.3–0.6	13.56 (g/m^2)	Lysozyme	3.18	0.4–2.0	805 (DBC) 1235 (SBC)
[38]	CA-DEAE and CA-COO	500	-	-	-	Lysozyme BSA	8.0/BSA5.5/ Lysozyme	2.0	COO-27 * lysozyme (DBC) & DEAE-20 * BSA (DBC)
[39]	PVA-MAH	226-(30% PVA) 284-(5% PVA)	45	0.002–0.064	3.2–3.5 (g/m^2)	Lysozyme	6.0	0.1–1.2	159 (DBC)
[40]	CMA	267	124	0.65	3.28 (g/m^2)	Lysozyme	4–8	1	160 (SBC) 118 (DBC)

 https://doi.org/10.3390/membranes16010005

Functional ligands are attached to the internal pore surfaces throughout the membrane structure and significantly enhance binding by increasing the number of accessible binding sites [12]. The grafting of charged or ion-exchange groups, such as carboxyl, DEAE, COO^-, and CCA, enhances electrostatic interactions between the membrane and target proteins, thereby improving binding efficiency [16,37,40]. This functionalization significantly improves the overall binding capacity, as the modified surface provides a higher density of accessible ligands for binding. For example, PAN-pAQ and PAN-GMA-DEA membranes functionalized with quaternary amine groups achieved up to 260 mg/g static binding capacity for BSA [35,36]. Notably, pH-responsive PVDF-g-PDMAEMA (poly(vinylidene fluoride)-graft-poly(dimethylaminoethyl methacrylate) membranes with anion exchangers achieved static and dynamic binding capacities of 318 mg/g and 210 mg/g, respectively [41]. Moreover, hydrophilic modification, as reflected by lower water contact angles, tends to promote protein accessibility and enhance binding. During functionalization, fibre diameter might vary, which can cause the loss of SSA because of the direct correlation between fibre diameter and SSA. Hence, variations in fibre diameter should be monitored before and after the functionalization of the nanofibrous membranes [1].

A smooth fibre surface and uniform fibre radius can be achieved after electrospinning the membrane, favouring higher throughput of the membrane during purification of proteins [42]. Tortuous micro channels also provide more available adsorption sites for the membranes [39]. For instance, the SA-PEO membrane, with fine nanofibrous networks and moderate pore sizes (0.3–0.6 μm), achieved exceptional lysozyme-binding capacities [37]. PVA-MAH membranes with different PVA content showed improved lysozyme adsorption due to their high specific surface area (3.2–3.5 g/m^2) and relatively low hydrophobicity, which together increase available binding sites and promote stronger protein–surface interactions [39]. These findings highlight that optimising nanofibre diameter, pore connectivity, and surface chemistry is crucial to balancing mass transfer and binding efficiency in next-generation ion-exchange membranes. In protein purification, high permeance typically requires membranes with porosity above 70%, and pores larger than 100 nm allow macromolecules to pass without affecting separation, while pore size directly influences binding capacity and permeability. Larger pores increase permeability but reduce specific surface area, whereas a narrow pore size distribution promotes uniform flow and enhances dynamic binding [1].

Future membrane fabrication for chromatographic applications should prioritise optimising nanofibrous membrane pore structures to enhance adsorption performance. Understanding pore architecture and surface modification effects on mass transport is essential for evaluating membrane-binding performance. Experimentally evaluating all parameters influencing process performance is time-consuming and resource-intensive. Computational simulations offer a powerful alternative by modelling flow and adsorption kinetics, providing insights into mass transport mechanisms and guiding membrane design. These simulations are particularly valuable for understanding how pore structure and surface properties influence the separation mechanism in ion-exchange processes, linking membrane characteristics directly to protein-binding efficiency.

2.1. Separation Mechanism of IEX Binding

In IEMC, the adsorption of protein on adsorbent membranes occurs via an ion-exchange mechanism. Separation and purification occur because of electrostatic interactions between the proteins and charged functional groups on the ion-exchange membrane [43]. Binding efficiency can be determined based on the nature of the proteins, adsorbents, and process conditions [44]. Ultimately, these factors affect the protein–adsorbent interactions, typically with electrostatic interactions being recognised as the primary influence. The electrostatic in-

 https://doi.org/10.3390/membranes16010005

teraction between the adsorbent (nanofibre surface) and the adsorbate (protein) plays a crucial role in protein binding. An adsorbent surface with a charge opposite to that of the protein can be more effective at adsorption compared to one with the same charge [29]. Additionally, the magnitude of the charges plays a crucial role in determining protein binding [40,45]. Therefore, for ion-exchange membranes, protein binding depends on the balance of electrostatic attraction between the surface charge of the membrane and the protein [46].

2.2. Factors Impacting the Binding of Proteins on IEX Nanofibres

The adsorption mechanism of IEX nanofibres is influenced by factors related to the surface of the adsorbent, protein characteristics, and solution conditions. The surface charge of the nanofibre, as determined by its zeta potential, directly affects the strength of protein attraction towards the surface. Similarly, the charge properties of the protein, governed by its pI (isoelectric point), the point of zero net charge, and the pH of the surrounding medium indicate whether the protein is negatively or positively charged [9]. Additionally, the ionic strength of the solution plays a critical role by influencing the Debye length, which determines the range of electrostatic forces [47]. Morphological properties related to both membranes, such as specific surface area (SSA), pore size, and porosity [35], and the protein, such as the size and shape [48], can also impact the magnitude of the electrostatic interactions. The charge of the protein can be altered by adjusting pH. When the pH is below the protein's isoelectric point (pI), the protein carries a net positive charge, whereas at pH values above the pI, the protein becomes net negative [49]. Since proteins exhibit varying electrical properties in buffer solutions at different pH levels, protein purification can be achieved by exploiting the controllable electrostatic interactions between the protein and the adsorbent [15].

2.2.1. Adsorbent Structural Effect

The ligand density can be correlated to the surface charge of the nanofibres [18]. Generally, as the ligand density increases, the binding capacity of the membrane also increases, exhibiting a direct and linear correlation due to the promoted grafting area [36,50]. On the other hand, the binding capacity was compromised with ligand density content as the adsorption capacity significantly reduced with a further increase in ligand content (after 3% w% of BTCA (butane tetracarboxylic acid)) because of a decrease in the effective adsorption area. This can refer to how the ligands are arranged or interact with each other, potentially blocking or obstructing available space for binding [51]. Similarly, the ligand density increased with pH (8.5 to 13), and IgG binding capacity increased from pH 8.5 to 10. However, despite the increasing ligand density from pH 10 to 13, the IgG binding capacity decreased during this range. This was linked to the limited accessibility of IgG to densely packed ligand binding sites in the grafted layers [52]. Optimum binding can be achieved by uniform ligand distribution on the IEX nanofibres [53]. Therefore, ligand density optimisation is important to employ the modification without compromising the membrane properties, such as porosity and fibre diameter. For instance, after functionalisation of the membrane, the porosity was compromised in the range of 1.14–4.6% [50], and in most cases, pore morphology was intact after the functionalization of the IEX membranes.

Besides grafting and modification, membrane and protein morphology are also crucial. For example, pore connectivity determines how effectively the internal pore network is interconnected, which directly affects intrapore diffusion and the accessibility of binding sites [54]. Highly interconnected pore networks facilitate uniform fluid pathways, thereby enhancing membrane performance, offering practical insights for the design of next-generation materials [55]. It can be assessed in a semi-quantitative manner through techniques such as tracer diffusion, permeability measurements, or 3D imaging approaches

 https://doi.org/10.3390/membranes16010005

(e.g., XCT, FIB-SEM) [56,57]. There is a complex trade-off relationship between surface micro/nano-structure and, hence, distribution and/or density of functional groups [18]. Overall, it was concluded that the high SSA of the nanofibrous IEX membrane allows a higher ligand density, thereby demonstrating a higher binding capacity. For instance, a higher adsorption capacity was reached with a 270% smaller fibre diameter and 400% higher SSA of the membrane in comparison to the study reported in [32]. Compared to the macroporous microfibres, IEX nanofibres had more binding capacity, which can underline the importance of the morphological properties of the membrane [18]. To understand the correlations between membrane structure and binding performance, various membranes were fabricated under varying electrospinning process parameters [35]. At a fixed pAQ content (15 mol%), the fibre orientation and alignment significantly influenced the protein binding capacity. Additionally, the binding with the well-oriented fibres increased by 45–66% compared to the randomly oriented fibres. Regarding protein morphology, studies found differences in binding to the membranes due to their charge differences [51]. Besides the isoelectric points and charge differences with the adsorbent surface, in some cases, the surface of the nanofibres was more favourable for specific types of proteins, indicating their distinctive nature for molecular weight [31,58].

The zeta potential of nanofibres and proteins plays an important role in ion-exchange chromatography. The overall surface charge, in any suspension, can be determined by the zeta potential values [59]. Zeta potential indicates the strength of surface charge on particles and primarily influences adsorption processes at longer time scales. The maximum adsorption was reached with the N-4ZH nanofibrous IEX membrane compared to the other membranes due to its highest zeta potential and uniform distribution of 4ZH nanoparticles within the fibres [29]. The effects of zeta potential can be explained by DLVO theory to describe the stability of the system by considering different forces acting on the particles [60]. High zeta potential, either positive or negative, is generally required to ensure stability. Thus, systems with zeta potentials $> \pm 30$ mV are generally considered stable. The DLVO approach has been successfully used for the evaluation of biological particle adhesion in the context of bioprocessing [61]. This approach was found as appropriate to be used as the theoretical framework for protein interaction studies to predict protein behaviour during binding [61]. With DLVO theory, protein interactions with the surface can be described as colloidal representations. The DLVO theory explains that the tendency of a colloidal particle to move toward or away from a surface depends on the balance of attraction and repulsion forces acting on it [62].

2.2.2. Protein Properties

Protein characteristics such as its structure (e.g., dimension, stability), size, and net charge affect the binding mechanism. The structure of proteins depends on the sequence of amino acids, and it can vary among proteins. The structural stability of proteins indicates the ease of undergoing conformational modification. Larger-sized proteins have more binding sites to interact with the surface, thereby increasing the possibility of binding to the surface. In certain instances, such as when the protein solution comprises a mixture of hundreds of different proteins, e.g., blood plasma, the mass transfer rate of protein molecules to the surface can be directly correlated with its concentration and inversely related to its molecular weight [63]. Through experimental observations, proteins that were at or near the isoelectric pH exhibited greater adsorption because of minimised charge–charge repulsion among the adsorbed molecules.

The adsorption behaviour of proteins is also controlled by the stability of a protein, which can be one of the driving forces during adsorption. The underlying reason for this is that the conformation of a folded protein is highly constrained, resulting in relatively

 https://doi.org/10.3390/membranes16010005

low entropy. However, if the protein undergoes partial or complete unfolding upon adsorption, this process can lead to an increase in conformational entropy, thereby acting as a driving force. Additionally, the conformational stability of a protein plays a critical role in determining the structure of the adsorbed protein. For example, soft proteins such as bovine serum albumin (BSA) can have relatively low structural stability and undergo extensive surface rearrangement. These can promote an increased number of interaction points among the proteins and the surface and adsorbed proteins. Hard proteins such as lysozyme and Chymotrypsin can be more resilient to major changes and can primarily adsorb to hydrophobic surfaces unless electrostatically attracted [64–66].

Electrostatic interactions in proteins are complex due to the heterogeneous medium surrounding protein charges. While water, the solvent in aqueous solutions, is highly polar with a dielectric constant of nearly 80 at room temperature, the protein interior is largely non-polar, with a dielectric constant typically ranging from 2 to 4, though values up to 40 have been reported. Charges within the protein interact with both the solvent and other charges, with the solvent effectively screening these charge–charge interactions. These charge–solvent interactions and the solvent's screening effect play a crucial role in shaping the electrostatic energies of proteins [67].

2.2.3. Solution Parameters

The pH value of the solution, as one of the crucial parameters for the performance of protein adsorption, is a determinative factor for the surface potentials of both nanofibres and protein molecules. While a lower pH may not be ideal for some proteins' charge magnitude for electrostatic interaction, it can be suitable for others. For example, pH 5.5 has the highest binding for IgG (pI = 7.8), whereas it can result in lower binding for BSA and lysozyme [42]. The reason behind this phenomenon is that when the pH is equivalent to 8.0, IgG is negatively charged, causing the electrostatic repulsion between the protein and the adsorption ligand. Conversely, the adsorption capacity of lysozyme increased as the pH decreased below the protein's pI 10.8, reaching a maximum at pH 6 due to the enhanced positive charge of lysozyme and negative charges on the PMA-functionalized nanofibres at lower pH [32]. For membranes with quaternary amine groups, the nanofibre surface charge remains positive and is substantially unchanged unless exposed to unrealistically extreme pH values. In contrast, membranes functionalised with tertiary amines show a strong pH-dependent surface charge, making pH the dominant factor governing their zeta potential [41]. As a result, electrostatic interactions can be controlled with a pH-responsive IEX membrane through switchable surface charges [68]. BSA adsorption was tuned by varying pH values in the range of 3.6–9.0. BSA binding increased sharply when the pH was increased from 3.6 to 6.4 and decreased sharply after pH 6.4. This is because at pH 3.6, the BSA molecule was positively charged, resulting in charge repulsion, and at pH 6.4 the BSA molecule (pI 4.7–4.9) was negatively charged, leading to significant binding of BSA. Additionally, a pH-controllable IEX nanofibrous membrane can be used for selective protein adsorption by regulating buffer pH value based on the pI of the target protein [34]. The effect of pH on BSA colloidal stability does not directly correlate with surface charge, as it reflects the interplay between van der Waals attraction and electric double-layer repulsion [69]. The variation in zeta potential with pH was more significant at pH values above 4.7 than at pH below 4.7 for BSA [70]. Therefore, it is important to understand the impact of pH on the electrostatic interaction between the IEX nanofibres and the protein.

The ionic strength of the solution can be adjusted to maximise the binding of proteins on IEX membranes. It can also be crucial during the elution of proteins by the electrostatic interactions through increasing the ionic strength [40]. The maximum IgG binding capacity

https://doi.org/10.3390/membranes16010005

was reached at the lowest ionic strength (0 M KCl), and IgG binding capacity decreased when the ionic strength increased [71]. This is because the Debye length decreases as ionic strength increases, leading to a screening of the electrostatic forces. The Debye length is inversely proportional to the square root of the ionic strength of the solution, which is determined by the concentration and the valency of the ions present [72,73]. The ionic radius of different ions, such as LiCl, NaCl, KCl, and $MgCl_2$, can affect the adsorption capacity of the membranes. With an increasing ionic radius, the adsorption capacity of lysozyme gradually decreased, possibly due to stronger shielding effects of ions with larger radii. A similar decrease in the adsorption capacity of the EVOH/BTCA nanofibrous IEX membrane was observed across the same ions [58]. The order of the adsorption capacities was LiCl > NaCl > KCl > $MgCl_2$, which is consistent with the findings reported in reference [74] for the same protein, lysozyme. Similarly, ions with larger radii caused stronger shielding effects, hence reducing the binding capacity. Another study highlighted that as the ionic strengths of buffer solutions significantly influences the electrostatic interaction between nanofibre and biomolecules, as the Na_2SO_4 concentration increased from 0.1 to 0.4 M, the adsorption capacity sharply decreased [31]. This resulted from the gradually enhanced shielding effect on the electrostatic interaction between the IEX nanofibre membrane and lysozyme molecules. Meanwhile, adsorption performance can be further tailored by adjusting buffer properties and protein concentrations. This emphasises the importance of identifying the optimal use for practical application.

3. Modelling Approaches for Protein Binding

Protein-binding models and computational tools vary substantially in their level of protein representation and in the transport phenomena they are designed to investigate. Due to the multi-scale nature of protein stability, it is computationally infeasible to employ a single model to address both nanometre-scale phenomena (e.g., conformational changes, protein–protein interactions) and macroscopic processes (e.g., aggregation, phase separation). Approaches to studying protein interactions can be classified as atomistic models, coarse-grain models, and continuum models [75]. Figure 1 illustrates these models and their time and length scales.

In atomistic models, each atom is represented as a single bead and an individual interaction site (Figure 2a). The physical properties (mass, charge, etc.) of atoms and the behaviour of the protein solution are described through the potential energy function [76]. This function incorporates bonded interactions such as bond stretching, angle bending, and intra- and intermolecular non-bonded interactions such as electrostatic, van der Waals forces, and hydrogen bonding across all molecular species. Molecular dynamics (MD) and Monte Carlo simulations can be used to give molecular insights into protein adsorption to understand the mechanisms and driving forces behind surface interactions, conformational changes, and adsorption kinetics at the atomic level [77]. Monte Carlo (MC) simulation employs repeated random sampling to produce numerical results and is particularly useful for systems with high degrees of freedom that are challenging to model using other methods. While Density Functional Theory (DFT) and molecular dynamics (MD) are widely applied, linking atomic-scale data to experimentally observable macroscopic properties remains a significant challenge. In such cases, coarse-grained (CG) and dissipative particle dynamics (DPD) methods are employed to investigate interactions at the mesoscale [48].

Coarse-grained models are used when the degrees of freedom of a molecule are reduced by grouping atoms into lumped/coarse-grained beads [78]. In addition to these models, hybrid approaches can be employed to integrate the strengths of atomistic and coarse-grained models while eliminating their limitations [78]. In CG models, larger time and length scales can be simulated by the incorporation of fewer interaction sites [79].

 https://doi.org/10.3390/membranes16010005

In DPD simulations, particles interact via a soft repulsive potential that permits overlap between non-bonded particles. Consequently, oppositely charged point species may readily form artificial ion pairs. To avoid this, a common approach is to smear the charges of DPD particles using spatial distributions [80]. Coarse-grained molecular models can also be defined on a lattice [81]. The Lattice Boltzmann Method (LBM) models fluid flow by tracking the evolution of particle distribution functions over a fixed spatial lattice, as governed by the discrete Boltzmann equations [82]. This method is a mesoscopic method that bridges the gap between continuum-based methods and molecular dynamics by modelling fluid behaviour at an intermediate scale [83]. Computational modelling becomes crucial as it can provide thorough molecular-level information at a scale that can be difficult to observe experimentally [84].

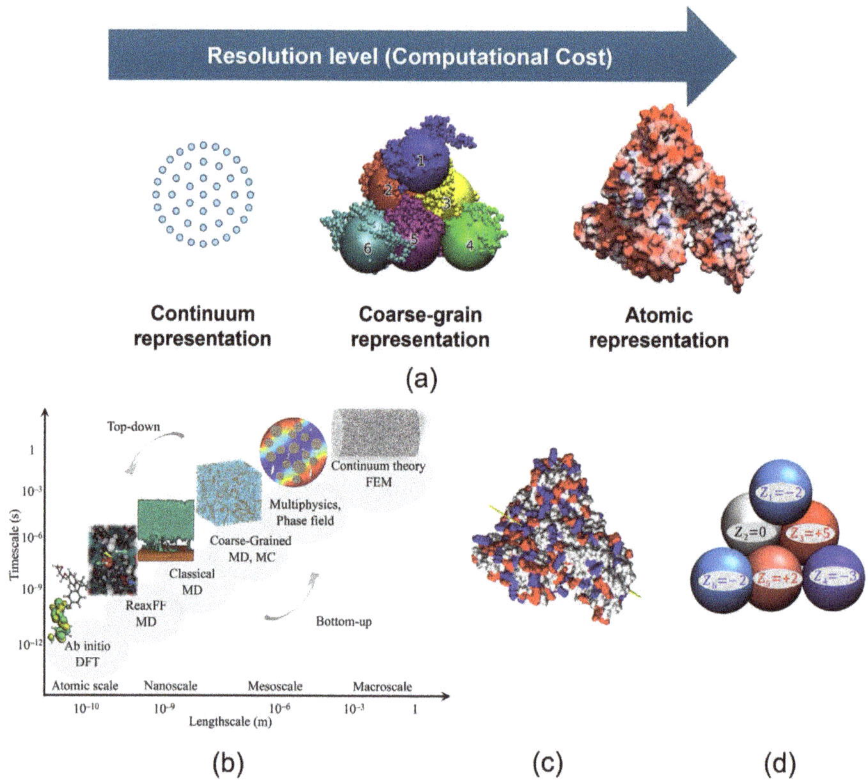

Figure 1. (**a**) Hierarchical classification of computational protein models by resolution level with BSA molecule; (**b**) a schematic representation of temporal and spatial scales; (**c**) distribution of charge on the BSA molecule surface coloured by residues' total charge: positive—blue, negative—red, neutral—white; (**d**) coarse-grained charge distribution of BSA at its isoelectric point, expressed in units of the elementary charge; *e* (redrawn from [76–79]) reproduced from Ref. [76] with permission from the Royal Society of Chemistry.

Continuum models offer a robust alternative for simulating protein adsorption that occurs on time and length scales beyond seconds and micrometres, which are inaccessible to atomistic and coarse-grained approaches [75]. The adsorption equilibrium of macromolecules on a porous material is generally represented using empirical or semi-empirical models, such as the Langmuir isotherm, the distributed pore model, or the Steric Mass Action (SMA) law,

https://doi.org/10.3390/membranes16010005

each offering varying levels of efficiency [85]. While these models can predict the salt effect, they can lead to unreliable results at high protein concentrations. Compared to other adsorption isotherms, the SMA model offers the advantage of explicitly incorporating counterion concentration while capturing both linear and nonlinear adsorption behaviours [86]. The SMA model employs stoichiometric binding theory by coupling the system's electrostatic and equilibrium characteristics within a set of correlation parameters [87]. Nonlinear adsorption behaviour up to adsorber saturation is explained by a reduction in available counter-ions on the adsorber surface. This occurs as adsorbed proteins either displace counter-ions or sterically shield them because of their size. Therefore, the model attributes adsorption limits not to the physical surface area of the adsorber but to its ionic capacity.

3.1. Modelling Frameworks for Ion-Exchange Protein Binding

Building on the existing protein-binding models, ion-exchange systems can also be analysed across different modelling scales, from molecular-level descriptions of electrostatic interactions to the macroscopic level used for parameter estimation and process design. Figure 2 shows the common adsorption models of protein adsorption on the IEX adsorber [88]. While stoichiometric adsorption models, such as the SMA model, offer a straightforward framework for describing protein adsorption, their capacity to accurately capture the electrostatic interactions between proteins and adsorbers is often debated [89]. Additionally, the SMA model requires prior information on three parameters for each macromolecule in solution: the characteristic charge, the equilibrium constant, and the steric factor. These parameters can be obtained through time- and resource-consuming chromatographic experiments involving gradient elution and breakthrough columns. As one of the challenges of these models, protein–protein interactions cannot be fully understood [90]. The theoretical limitations of stoichiometric models in describing electrostatic interactions between charged proteins and the charged adsorber surface, particularly concerning ionic strength and pH, have led to the creation of non-stoichiometric adsorption models [75].

CFD has been used to examine how membrane structural features, such as channel tortuosity, pore size, and connectivity and device design, have a significant influence on fluid flow distribution, as poorly engineered configurations can lead to substantial flow maldistribution that diminishes binding performance [55,91]. CFD tools enable the optimisation of membrane structures by modelling fluid dynamics, thereby informing the design of architectures that promote improved flow uniformity and performance. CFD simulations offer high-resolution flow fields around fibres and can be integrated with particle dynamics models, such as the discrete phase model (DPM) and discrete element model (DEM), to enable detailed analysis of particle permeation and behaviour near the fibres [92]. With CFD–DPM (discrete phase model) simulations, it is possible to track particles within the fluid medium and to study deposition behaviour of these sub-micron particles [93]. The CFD–DEM simulations predict colloid trajectories by computing the net force on each particle, incorporating hydraulic forces from the surrounding fluid flow, which can be gained through coupling with CFD [94]. Thus, using CFD coupled with DEM (discrete element method) would enable us to study each particle interaction with the surface as well as with the other particles and the surrounding medium [95]. Also, DEM simulations can simulate the particle interaction in the range of 10^6–10^8 particles [96]. The population balance model (PBM) simulations can predict the particle size changes in bulk-level growth to simulate particle aggregation during binding [97]. CFD–PBM (population balance model) simulations enable the evaluation of population dynamics governed by nucleation, growth, dispersion, dissolution, aggregation, and breakage by studying particle size distribution (PSD) on surfaces [98]. PBM and DEM are capable

 https://doi.org/10.3390/membranes16010005

of capturing aggregation dynamics of particles, with PBM focusing on population-level characteristics through statistical approaches and DEM resolving particle-level interaction in detail [99]. In contrast, DPM cannot provide key information such as collision frequency and aggregation effects, limiting its suitability for studying particle aggregation [100]. Additionally, in DPM, the particles are treated as a dimensionless point mass; therefore, monitoring the distance between the surfaces of the particle and fibre is necessary during the trajectories [101]. While particle–particle (protein–protein) interactions during binding can be studied with DEM, DPM, and PBM simulations, it is not possible to study these interactions, which can limit their suitability for IEX binding prediction fully.

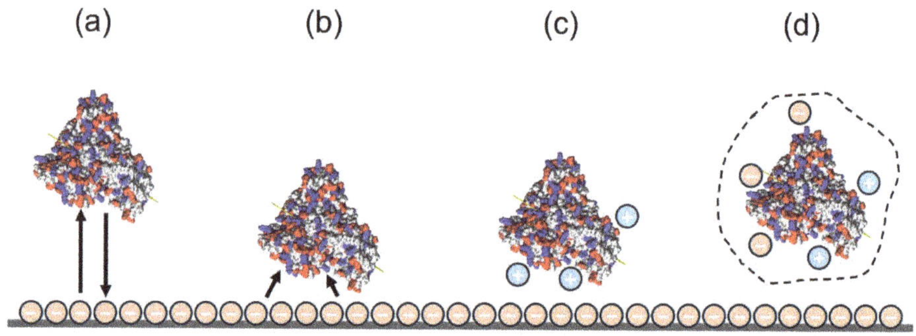

Figure 2. Commonly used binding models in ion-exchange: (**a**) the Langmuir model; (**b**) the stoichiometric model; (**c**) the Steric Mass Action (SMA) model; (**d**) the colloidal particle adsorption model. The BSA molecule is shown with its surface charge distribution coloured by the total charge of individual residues (blue: positive, red: negative, white: neutral). The "+" and "−" symbols represent mobile ions in the electrolyte surrounding the protein; the arrows illustrate electrostatic interactions with the surface under each modelling framework (redrawn from Ref. [92]).

To improve the IEX binding and to design new high-throughput adsorptive membranes, the adsorption process needs to be understood. However, since the process occurs on a molecular level, it is not possible to observe the process through experimental techniques. These alternative models offer a more fundamental understanding of electrostatic interactions in ion-exchange (IEX) chromatography by using an idealised colloidal representation of the protein and analytical solutions to the linear Poisson–Boltzmann equation [102]. Theoretical modelling is required to obtain precise analytical expressions for the interaction potential essential for deducing nanoparticle properties such as size or charge and for computing phase behaviour or transport coefficients [48].

3.2. Application of DLVO Theory to Protein Binding on IEX Nanofibres

Colloidal interactions are often described by the DLVO theory, which is named after Derjaguin and Landau and Verwey and Overbeek [103]. DLVO theory offers a conceptually straightforward and accurate framework for describing interactions between surfaces separated by a liquid, playing a crucial role in understanding colloidal system behaviour [104]. DLVO theory attributes surface interactions to the combined effects of van der Waals forces and electrostatic double-layer interactions (Figure 3a).

At the primary minimum, interparticle interaction is strongest, leading to an unstable suspension characterised by strong aggregation or irreversible aggregation. At the energy barrier, repulsive forces dominate, resulting in a stable dispersion. The secondary minimum reflects weaker attraction, allowing for reversible or weak aggregation (Figure 3b) [105]. The energy barrier height increases with decreasing ionic strength, increasing zeta potential,

 https://doi.org/10.3390/membranes16010005

and larger particle size. Conversely, the secondary minimum increases with higher ionic strength and particle size but lower zeta potential [60]. The DLVO model was used to predict the structural behaviour of lysozyme solutions under various conditions of pH and ionic strength [106]. It was found that DLVO theory generally captures the trend of protein interactions becoming increasingly short-range and attractive with rising pH and ionic strength. The DLVO theory also used another study to describe the interaction between BSA and PES using zeta potential, surface charge density, and interaction energy [107]. In the study, a secondary minimum of the interaction energy governed the binding of BSA on PES. Additionally, the interaction energy decreased with increasing ionic strength, while adsorption increased as the system pH decreased.

Besides DLVO theory, extended DLVO (xDLVO) theory was also used through the surface energetics approach to explain the interaction behaviour of proteins with IEX chromatographic supports [59]. According to xDLVO theory, protein interactions are governed by the combined effects of Lewis acid-base, van der Waals, and coulombic type forces [108]. xDLVO theory was used to explore the parameters used for the surface energetics approach to calculate the interaction energy minimum of different proteins binding onto three different supports, e.g., glass, plastic, and membrane [108]. The interaction behaviour is significantly influenced by ligand type (phenyl vs. butyl), ligand density (25–75%), and lambda conditions (0.2–1.0 nm). The surface energy components can be calculated from two experimental measurements: the contact angle, which helps to determine the Lifshitz van der Waals and acid-base components, and the zeta potential, which is used to determine the electrostatic component of surface energy [59].

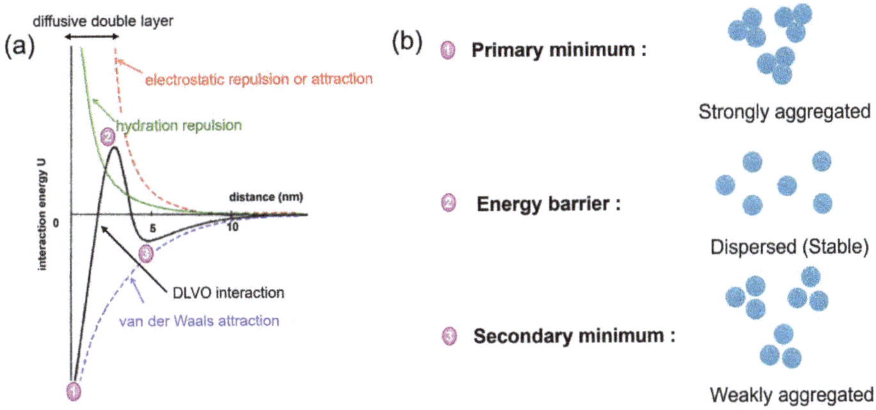

Figure 3. (**a**) The occurrence of three interaction forces: when a protein molecule approaches a surface, electrostatic attraction or repulsion, van der Waals attraction, and hydration repulsion; (**b**) description of the stability of colloidal particles (redrawn from Ref. [109]).

3.3. Dynamics of Fibre–Protein Interactions

Understanding the interactions of proteins with the surfaces is based on the interaction energies between the proteins and the fibre surface. Interaction energy calculations can be commonly used to predict situations as either favourable or unfavourable for colloid aggregation. Protein binding on a surface is a complex process influenced by (1) protein properties, e.g., conformation, charge distribution, the strength of intramolecular bonds; (2) surface properties, e.g., chemistry, charge, roughness, surface energy; and (3) solution conditions, e.g., pH, salt concentration [109,110]. Because of these factors, the interactions are considered dynamic processes that theoretically involve several potentially interchangeable steps [111].

 https://doi.org/10.3390/membranes16010005

Surface morphology, besides the electrostatic interactions, can alter protein binding on surfaces [112]. The interaction between the colloid and surface is highly dependent on the nanoscale surface roughness properties of both surfaces. The influence of nanoscale roughness is likely to change with the solution chemistry, the colloid size, the surface charges, and the water velocity [111]. Roughness is a term used to characterise surface irregularities. The artificial modification of surface roughness was achieved by increasing the friction coefficient [113]. The interactions between particles and fibres were described in terms of friction and restitution coefficients. Friction between particle and fibre was found to influence particle behaviour during impact and adhesion [114]. The increase in the particle–wall friction coefficient led to an increase in the charge magnitude of the surface, as this allows a longer contact duration with the surface, causing more opportunities for charge transfer [115]. Most colloidal particles in suspension (0.001–10 μm) tend to remain suspended and settle very slowly in solution to maintain electrical neutrality due to the presence of surface charge arising from isomorphic substitution, chemical reaction at the interface, or preferential adsorption of ions on the particle surface from the surroundings by excess of oppositely charged ions (counterions). The combined interactions of this system of oppositely charged clouds are known as the electric double layer. The developed electrostatic repulsive forces prevent particle aggregation in the colloidal system and, hence, contribute to their stability in the dispersed system [116].

Surface energy is another important property that influences the interaction of a protein with the surface. Since proteins are charged molecules, a charged surface will impact their interactions [117]. At uncharged fluid interfaces, hydrophobic interactions primarily govern the binding behaviour, whereas at charged surfaces, a combination of hydrophobic and electrostatic interactions becomes more significant [84]. Hydrophobic surfaces have the advantage of interacting with hydrophobic residues of proteins, and water displacement from the surface is more favourable than on hydrophilic surfaces. On the other hand, the hydrophilic membranes were generally found to be more advantageous due to their possibility of reducing non-specific binding [1]. Wettability of surfaces not only influences the total amount of proteins but also their spatial conformation. For example, it was shown that the impact of surface wettability on protein confirmation during adsorption for bovine serum albumin (BSA) demonstrates that proteins adsorbed on more hydrophobic materials present a higher degree of denaturation. Usually, hydrophobic materials are non-polar, while hydrophilic substrates present a distribution of charges on their surfaces [118].

3.4. *Influence of Protein–Protein Interactions During Binding*

Protein–protein interactions can lead to challenges, such as aggregation, liquid–liquid phase separation, and increased solution viscosity, particularly at high protein concentrations [119]. The aggregation tendency of proteins can be minimised by controlling several factors such as temperature, pH, ionic strength, and protein concentration. When proteins are more likely to form partially folded states that lead to aggregation, a high net charge reduces their tendency to aggregate. This can cause repulsive double-layer forces to increase the colloidal stability of the protein. In solutions close to the protein's isoelectric pH (pI), strong and attractive electrostatic interactions between proteins have been associated with enhanced aggregation tendencies [120]. Proteins at their pI lack net charge, weakening electrostatic repulsion [119]. The reduction in the net energy barrier for protein–protein interaction led to a loss of stability, thereby enhancing aggregation [121]. The aggregate binding mechanism is predominantly driven by electrostatic interactions [122].

Colloidal protein–protein interactions, including attractive and repulsive nature, can be decisive in protein's solubility, aggregation, precipitation, and crystallisation. These interac-

 https://doi.org/10.3390/membranes16010005

tions alone were not possible control or interfere with through observations or examinations during the experiments. Hence, protein–protein interactions are commonly quantified using the osmotic second virial coefficient (B_{22}) or the diffusion interaction parameter (k_D) [120]. There are a few models, such as Monte Carlo and molecular dynamics, that can be used to estimate B_{22} [123]. Among these models, DLVO theory can also be used to calculate B_{22} coefficients to better understand protein–protein interactions to optimise process conditions towards the stability of proteins [124]. Normally, the extended DLVO (xDLVO) model is used to describe B_{22} in aqueous solution as a function of pH, salt type and concentration, and temperature by fitting to experimental B_{22} data [124]. There are available coarse-grained simulations to quantify the binding of proteins on the surfaces. Since the hardship of quantification of such interactions, extended DLVO theory (xDLVO) can be used with a coarse-grained model (xDLVO-CG) to extend the existing models by a coarse-grained representation of protein and the inclusive additional ion–protein dispersion interaction term [124].

Generally, the adsorption of protein increases with the surface charge, but if the surface charge is too high, protein adsorption decreases due to the flat orientation of the protein. When increasing the charge of the surface, the adsorbed mass of a given charged protein increases. However, the adsorption of protein is very low if the charge on the surface is very high, probably due to the flat orientation of the protein [125]. A simplified protein–protein interaction model was based on a Baxter adhesive potential and an electric double-layer force to distinguish the contributions of longer-ranged electrostatic interactions from short-ranged attractive forces. The DLVO theory was successfully adopted to accurately capture the dependence of the ionic strength for solutions at pH 6.5 and below [119].

Multilayer adsorption is dependent on both the protein type and concentration. For example, lysozyme binds to the surface due to its strong positive surface charge, showing higher electrostatic interaction than α-chymotrpsin. In contrast, α-chymotrpsin can form multilayer binding in both low and high concentrations, facilitated by its heterogeneous charge distribution and limited surface area of adsorptive surface availability [85]. Protein deposition on the surface of the fibre could change the dynamics of the adsorption and corresponding governing equations, as shown in Figure 4.

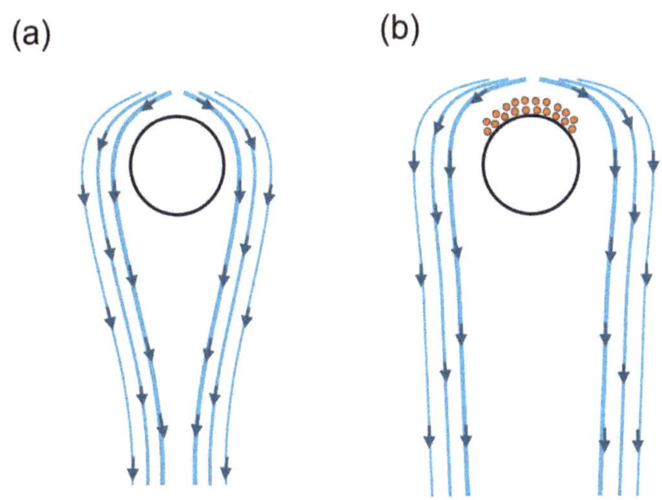

Figure 4. (**a**) Schematic representation of flow field around the fibre without discrete particles; (**b**) changed flow field due to deposited/bound particles on the fibre. The arrows represent the flow direction around the fibre. (redrawn from Ref. [121]).

 https://doi.org/10.3390/membranes16010005

Therefore, it is crucial to simulate the protein–protein interactions along with the protein–surface interaction for studying the adsorption of proteins on electrospun nanofibres. Binding energies can be studied in detail while accounting for van der Waals and electrostatic repulsion (DLVO) forces between the proteins and the fibre surface. Additionally, the relationship between binding and membrane properties can also be studied in detail.

4. Conclusions and Future Perspectives

Colloidal and meso-scale modelling approaches provide a powerful tool "sweet spot" for analysing protein binding mechanisms, offering sufficient resolution to capture key interaction dynamics with no need for the computational cost of fully atomistic simulations. These methods bridge molecular interactions with macroscopic binding properties enabling detailed visualisation and quantification of how pore geometry, surface charge distribution, and local hydrodynamics influence protein transport and binding within nanofibrous or porous membranes.

Protein binding is influenced by numerous factors, including adsorbent properties, protein characteristics, and solution conditions. Experimentally evaluating each of these individually can be time-consuming and resource-intensive. Therefore, computational approaches allow systematic exploration of these parameters, providing predictive insights into their effects on binding behaviour. When combined with theoretical frameworks such as DLVO-based interaction energy calculations, they offer predictive insight into protein–fibre interactions, binding propensity, and competitive adsorption effects under different solution conditions. By capturing interplay between electrostatics, steric constraints, and convective transport, such models directly support rational membrane design by identifying optimal surface chemistries, pore architectures, and operating conditions required to maximise binding efficiency, selectivity, and overall chromatographic performance.

Future membrane fabrication for chromatographic applications should prioritise optimising the pore structures of nanofibrous membranes to enhance adsorption performance. Understanding pore architecture and surface modification effects on mass transport is essential for evaluating membrane binding performance. Overall, the concepts and modelling frameworks summarised in this work not only enhance our understanding of protein binding in membrane chromatography but also extend to broader applications in the design of protein purification and antifouling membranes, where understanding protein–surface interactions is crucial. This broader relevance highlights the wider impact of the modelling frameworks and their potential to support future advances across membrane-based separation technologies. Future research should extend these modelling frameworks to more accurately represent flexible nanofibre systems, where dynamic processes can be captured. Adopting colloidal-scale approximations of proteins offers a computationally efficient yet physically meaningful route to study binding dynamics on nanofibrous materials. By representing proteins as soft colloidal particles, such models can capture essential features of adsorption and transport without the prohibitive cost of fully atomistic simulations. This approach enables systematic exploration of critical process parameters such as surface charge density, pore connectivity, and flow conditions that govern binding performance. Consequently, colloidal-level modelling can accelerate and guide the design of high-throughput and cost-effective ion-exchange membranes by providing insights that bridge molecular detail with process-scale functionality.

The future of IEX-NFM design lies in the convergence of physics-based modelling and data-driven intelligence. Machine learning models trained on experimental and simulated datasets could rapidly predict protein-binding capacity, pressure drop, and mass transport

https://doi.org/10.3390/membranes16010005

behaviour from structural inputs such as fibre morphology and properties [126]. Coupled with material informatics approaches for feature selection and design, this will allow accelerated optimisation of membrane architectures. In the long run, integrating AI-guided design and high-throughput characterisation could establish continuous improvement of membrane materials [127].

Author Contributions: Conceptualization, I.B., X.Y., and L.K.; methodology, I.B.; validation, I.B., S.Z., X.Y., and L.K.; formal analysis, I.B., S.Z., X.Y., and L.K.; investigation, I.B., S.Z., X.Y., and L.K.; resources, L.K.; data curation, I.B.; writing—original draft preparation, I.B.; writing—review and editing, I.B., S.Z., X.Y., and L.K.; supervision, S.Z., X.Y., and L.K.; project administration, L.K.; funding acquisition, L.K. All authors have read and agreed to the published version of the manuscript.

Funding: This research was funded by the Australian Research Council (ARC) Discovery scheme (DP220102729 for L.K. and DP230100307 for S.Z.) and Research Foundation-Flanders (FWO) Odysseus Grant (#G0F7621N for X.Y.).

Institutional Review Board Statement: Not applicable.

Informed Consent Statement: Not applicable.

Data Availability Statement: No new data were created or analyzed in this study. Data sharing is not applicable to this article.

Acknowledgments: This work was performed in part at the Deakin node and the Melbourne Centre for Nanofabrication (MCN) of the Australian National Fabrication Facility (ANFF).

Conflicts of Interest: The authors declare no conflicts of interest.

References

1. Yang, X.; Merenda, A.; Al-Attabi, R.; Dumée, L.F.; Zhang, X.; Thang, S.H.; Pham, H.; Kong, L. Towards next generation high throughput ion exchange membranes for downstream bioprocessing: A review. *J. Membr. Sci.* **2022**, *647*, 120325. [CrossRef]
2. Zydney, A.L. New developments in membranes for bioprocessing—A review. *J. Membr. Sci.* **2021**, *620*, 118804. [CrossRef]
3. Janson, J.-C.; Rydén, L. *Protein Purification: Principles, High Resolution Methods, and Applications*; John Wiley & Sons: Hoboken, NJ, USA, 1989.
4. Nikolaos, E.L. *Handbook on Protein Purification: Industry Challenges and Technological Developments*; Nova: New York, NY, USA, 2018.
5. Gerstweiler, L.; Bi, J.; Middelberg, A.P.J. Continuous downstream bioprocessing for intensified manufacture of biopharmaceuticals and antibodies. *Chem. Eng. Sci.* **2021**, *231*, 116272. [CrossRef]
6. Murphy, C.; Devine, T.; O'Kennedy, R. Technology advancements in antibody purification. *Antib. Technol. J.* **2016**, *6*, 17–32.
7. Liu, S.; Li, Z.; Yu, B.; Wang, S.; Shen, Y.; Cong, H. Recent advances on protein separation and purification methods. *Adv. Colloid Interface Sci.* **2020**, *284*, 102254. [CrossRef] [PubMed]
8. Chen, J.; Yu, B.; Cong, H.; Shen, Y. Recent development and application of membrane chromatography. *Anal. Bioanal. Chem.* **2023**, *415*, 45–65. [CrossRef] [PubMed]
9. Li, Y.; Stern, D.; Lock, L.L.; Mills, J.; Ou, S.-H.; Morrow, M.; Xu, X.; Ghose, S.; Li, Z.J.; Cui, H. Emerging biomaterials for downstream manufacturing of therapeutic proteins. *Acta Biomater.* **2019**, *95*, 73–90. [CrossRef] [PubMed]
10. Saufi, S.M.; Fee, C.J. Mixed matrix membrane chromatography based on hydrophobic interaction for whey protein fractionation. *J. Membr. Sci.* **2013**, *444*, 157–163. [CrossRef]
11. Ye, J.; Chu, J.; Yin, J.; Zhang, Y.; Meng, J. Surface modification of regenerated cellulose membrane based on thiolactone chemistry—A novel platform for mixed mode membrane adsorbers. *Appl. Surf. Sci.* **2020**, *511*, 145539. [CrossRef]
12. Liu, H.F.; Ma, J.; Winter, C.; Bayer, R. Recovery and purification process development for monoclonal antibody production. *mAbs* **2010**, *2*, 480–499. [CrossRef]
13. Francis, P. Modeling of Controlled-Shear Affinity Filtration Using Computational Fluid Dynamics and a Novel Zonal Rate Model for Membrane Chromatography. Doctoral Dissertation, University of British Columbia, Vancouver, BC, Canada, 2011.
14. Lalli, E.; Silva, J.S.; Boi, C.; Sarti, G.C. Affinity Membranes and Monoliths for Protein Purification. *Membranes* **2020**, *10*, 1. [CrossRef] [PubMed]

 https://doi.org/10.3390/membranes16010005

15. Fu, Q.; Duan, C.; Yan, Z.; Si, Y.; Liu, L.; Yu, J.; Ding, B. Electrospun nanofibrous composite materials: A versatile platform for high efficiency protein adsorption and separation. *Compos. Commun.* **2018**, *8*, 92–100. [CrossRef]
16. Fu, Q.; Wang, X.; Si, Y.; Liu, L.; Yu, J.; Ding, B. Scalable Fabrication of Electrospun Nanofibrous Membranes Functionalized with Citric Acid for High-Performance Protein Adsorption. *ACS Appl. Mater. Interfaces* **2016**, *8*, 11819–11829. [CrossRef]
17. Zobel-Roos, S.; Stein, D.; Strube, J. Evaluation of Continuous Membrane Chromatography Concepts with an Enhanced Process Simulation Approach. *Antibodies* **2018**, *7*, 13. [CrossRef]
18. Yang, X.; Hsia, T.; Merenda, A.; Al-Attabi, R.; Dumee, L.F.; Thang, S.H.; Kong, L. Constructing novel nanofibrous polyacrylonitrile (PAN)-based anion exchange membrane adsorber for protein separation. *Sep. Purif. Technol.* **2022**, *285*, 120364. [CrossRef]
19. Rathore, A.S.; Kumar, D.; Kateja, N. Recent developments in chromatographic purification of biopharmaceuticals. *Biotechnol. Lett* **2018**, *40*, 895–905. [CrossRef]
20. Borneman, Z.; Groothuis, B.; Willemsen, M.; Wessling, M. Coiled fiber membrane chromatography. *J. Membr. Sci.* **2010**, *346*, 327–334. [CrossRef]
21. Boi, C.; Malavasi, A.; Carbonell, R.G.; Gilleskie, G. A direct comparison between membrane adsorber and packed column chromatography performance. *J. Chromatogr. A* **2020**, *1612*, 460629. [CrossRef] [PubMed]
22. Roque, A.C.A.; Pina, A.S.; Azevedo, A.M.; Aires-Barros, R.; Jungbauer, A.; Di Profio, G.; Heng, J.Y.Y.; Haigh, J.; Ottens, M. Anything but Conventional Chromatography Approaches in Bioseparation. *Biotechnol. J.* **2020**, *15*, 1900274. [CrossRef] [PubMed]
23. Ramos-de-la-Peña, A.M.; González-Valdez, J.; Aguilar, O. Protein A chromatography: Challenges and progress in the purification of monoclonal antibodies. *J. Sep. Sci.* **2019**, *42*, 1816–1827. [CrossRef]
24. Wang, X.; Cheng, P.; Liu, N.; Wan, Y.; Guo, Q.; Cheng, Q.; Liu, K.; Lu, Z.; Li, M.; Wang, D. Highly efficient nanofibrous sterile membrane with anti-BSA/RNA-fouling surface via plasma-assisted carboxylation process. *J. Membr. Sci.* **2020**, *601*, 117935. [CrossRef]
25. Qiao, L.; Lei, S.; Du, K. High-surface-area interconnected macroporous nanofibrous cellulose microspheres: A versatile platform for large capacity and high-throughput protein separation. *Cellulose* **2021**, *28*, 2125–2136. [CrossRef]
26. Zhang, S.; Tanioka, A.; Matsumoto, H. De Novo Ion-Exchange Membranes Based on Nanofibers. *Membranes* **2021**, *11*, 652. [CrossRef]
27. Lavoie, J.; Fan, J.; Pourdeyhimi, B.; Boi, C.; Carbonell, R.G. Advances in high-throughput, high-capacity nonwoven membranes for chromatography in downstream processing: A review. *Biotechnol. Bioeng.* **2024**, *121*, 2300–2317. [CrossRef]
28. Dabaghian, Z.; Pakkaner, E.; Kong, L.; Yang, X. A review on chromatographic membranes for antibody purification: Factors affecting binding capacity and guidelines for membrane design. *Sep. Purif. Technol.* **2025**, *357*, 130147. [CrossRef]
29. Esfahani, H.; Prabhakaran, M.P.; Salahi, E.; Tayebifard, A.; Rahimipour, M.R.; Keyanpour-Rad, M.; Ramakrishna, S. Electrospun nylon 6/zinc doped hydroxyapatite membrane for protein separation: Mechanism of fouling and blocking model. *Mater. Sci. Eng. C* **2016**, *59*, 420–428. [CrossRef]
30. Agarwal, S.; Greiner, A. On the way to clean and safe electrospinning—Green electrospinning: Emulsion and suspension electrospinning. *Polym. Adv. Technol.* **2011**, *22*, 372–378. [CrossRef]
31. Lv, H.; Wang, X.; Fu, Q.; Si, Y.; Yin, X.; Li, X.; Sun, G.; Yu, J.; Ding, B. A versatile method for fabricating ion-exchange hydrogel nanofibrous membranes with superb biomolecule adsorption and separation properties. *J. Colloid Interface Sci.* **2017**, *506*, 442–451. [CrossRef] [PubMed]
32. Najafi, M.; Chery, J.; Frey, M.M. Functionalized Electrospun Poly(Vinyl Alcohol) Nanofibrous Membranes with Poly(Methyl Vinyl Ether-Alt-Maleic Anhydride) for Protein Adsorption. *Materials* **2018**, *11*, 1002. [CrossRef] [PubMed]
33. Hu, J.; Kumar, B.; Lu, J. *Handbook of Fibrous Materials, 2 Volumes: Volume 1: Production and Characterization/Volume 2: Applications in Energy, Environmental Science and Healthcare*; John Wiley & Sons: Hoboken, NJ, USA, 2020.
34. Li, G.; Li, T.; Li, Y.; An, L.; Li, W.; Zhang, Z. Preparation of pH-controllable nanofibrous membrane functionalized with lysine for selective adsorption of protein. *Colloids Surf. A Physicochem. Eng. Asp.* **2017**, *531*, 173–181. [CrossRef]
35. Al-Attabi, R.; Merenda, A.; Hsia, T.; Sriramoju, B.; Dumée, L.F.; Thang, S.H.; Pham, H.; Yang, X.; Kong, L. Morphology engineering of nanofibrous poly(acrylonitrile)-based strong anion exchange membranes for enhanced protein adsorption and recovery. *J. Water Process Eng.* **2024**, *65*, 105750. [CrossRef]
36. Chen, S.-T.; Wickramasinghe, S.R.; Qian, X. Electrospun Weak Anion-Exchange Fibrous Membranes for Protein Purification. *Membranes* **2020**, *10*, 39. [CrossRef]
37. Dou, X.; Wang, Q.; Li, Z.; Ju, J.; Wang, S.; Hao, L.; Sui, K.; Xia, Y.; Tan, Y. Seaweed-Derived Electrospun Nanofibrous Membranes for Ultrahigh Protein Adsorption. *Adv. Funct. Mater.* **2019**, *29*, 1905610. [CrossRef]
38. Dods, S.R.; Hardick, O.; Stevens, B.; Bracewell, D.G. Fabricating electrospun cellulose nanofibre adsorbents for ion-exchange chromatography. *J. Chromatogr. A* **2015**, *1376*, 74–83. [CrossRef]

 https://doi.org/10.3390/membranes16010005

39. Wang, X.; Fu, Q.; Wang, X.; Si, Y.; Yu, J.; Wang, X.; Ding, B. In situ cross-linked and highly carboxylated poly(vinyl alcohol) nanofibrous membranes for efficient adsorption of proteins. *J. Mater. Chem. B* **2015**, *3*, 7281–7290. [CrossRef]
40. Ma, J.; Wang, X.; Fu, Q.; Si, Y.; Yu, J.; Ding, B. Highly Carbonylated Cellulose Nanofibrous Membranes Utilizing Maleic Anhydride Grafting for Efficient Lysozyme Adsorption. *ACS Appl. Mater. Interfaces* **2015**, *7*, 15658–15666. [CrossRef]
41. Gao, P.; Huang, J.; Pakkaner, E.; Wagemans, J.; Eyley, S.; Thielemans, W.; Gijsbers, R.; Smet, M.; Yang, X. Tailoring stimuli-responsive PVDF-based copolymer membrane with engineered pore structure for efficient antibody purification. *Chem. Eng. J.* **2023**, *476*, 146700. [CrossRef]
42. Ye, J.; Wang, X.; Chu, J.; Yao, D.; Zhang, Y.; Meng, J. Electrospun poly(styrene-co-maleic anhydride) nanofibrous membrane: A versatile platform for mixed mode membrane adsorbers. *Appl. Surf. Sci.* **2019**, *484*, 62–71. [CrossRef]
43. Liang, J.; Fieg, G.; Keil, F.J.; Jakobtorweihen, S. Adsorption of Proteins onto Ion-Exchange Chromatographic Media: A Molecular Dynamics Study. *Ind. Eng. Chem. Res.* **2012**, *51*, 16049–16058. [CrossRef]
44. Yao, Y.; Lenhoff, A.M. Electrostatic Contributions to Protein Retention in Ion-Exchange Chromatography. 1. Cytochrome *c* Variants. *Anal. Chem.* **2004**, *76*, 6743–6752. [CrossRef]
45. Esfahani, H.; Prabhakaran, M.P.; Salahi, E.; Tayebifard, A.; Keyanpour-Rad, M.; Rahimipour, M.R.; Ramakrishna, S. Protein adsorption on electrospun zinc doped hydroxyapatite containing nylon 6 membrane: Kinetics and isotherm. *J. Colloid Interface Sci.* **2015**, *443*, 143–152. [CrossRef]
46. Chen, J.; Peng, R.; Chen, X. Hydrophobic interaction membrane chromatography for bioseparation and responsive polymer ligands involved. *Front. Mater. Sci.* **2017**, *11*, 197–214. [CrossRef]
47. Rabe, M.; Verdes, D.; Seeger, S. Understanding protein adsorption phenomena at solid surfaces. *Adv. Colloid Interface Sci.* **2011**, *162*, 87–106. [CrossRef]
48. Dahirel, V.; Jardat, M. Effective interactions between charged nanoparticles in water: What is left from the DLVO theory? *Curr. Opin. Colloid Interface Sci.* **2010**, *15*, 2–7. [CrossRef]
49. Umatheva, U. Computational Fluid Dynamics Simulations of Membrane and Resin-Based Chromatography. Doctoral Dissertation, McMaster University, Hamilton, ON, Canada, 2019.
50. Lemma, S.M.; Boi, C.; Carbonell, R.G. Nonwoven Ion-Exchange Membranes with High Protein Binding Capacity for Bioseparations. *Membranes* **2021**, *11*, 181. [CrossRef] [PubMed]
51. Fu, Q.; Si, Y.; Liu, L.; Yu, J.; Ding, B. Elaborate design of ethylene vinyl alcohol (EVAL) nanofiber-based chromatographic media for highly efficient adsorption and extraction of proteins. *J. Colloid Interface Sci.* **2019**, *555*, 11–21. [CrossRef]
52. Fan, J.; Boi, C.; Lemma, S.M.; Lavoie, J.; Carbonell, R.G. Iminodiacetic Acid (IDA) Cation-Exchange Nonwoven Membranes for Efficient Capture of Antibodies and Antibody Fragments. *Membranes* **2021**, *11*, 530. [CrossRef]
53. Rajesh, S.; Schneiderman, S.; Crandall, C.; Fong, H.; Menkhaus, T.J. Synthesis of Cellulose-graft-Polypropionic Acid Nanofiber Cation-Exchange Membrane Adsorbers for High-Efficiency Separations. *ACS Appl. Mater. Interfaces* **2017**, *9*, 41055–41065. [CrossRef]
54. Wang, J.; Dismer, F.; Hubbuch, J.; Ulbricht, M. Detailed analysis of membrane adsorber pore structure and protein binding by advanced microscopy. *J. Membr. Sci.* **2008**, *320*, 456–467. [CrossRef]
55. Yang, X.; Tao, L.; Li, Y.; Liu, Q.; Zhou, H.; Wan, Y.; Chen, G. The Effect of Pore Channel Structure Uniformity on the Performance of a Membrane Adsorber Determined by Flow Distribution Analysis Using Computational Fluid Dynamics. *Processes* **2025**, *13*, 956. [CrossRef]
56. Fallahianbijan, F.; Giglia, S.; Carbrello, C.; Zydney, A.L. Quantitative analysis of internal flow distribution and pore interconnectivity within asymmetric virus filtration membranes. *J. Membr. Sci.* **2020**, *595*, 117578. [CrossRef]
57. Alsoy Altinkaya, S. A review on microfiltration membranes: Fabrication, physical morphology, and fouling characterization techniques. *Front. Membr. Sci. Technol.* **2024**, *3*. [CrossRef]
58. Jiang, Y.; Lu, J.; Guo, L. Fabrication of highly carboxylated thermoplastic nanofibrous membranes for efficient absorption and separation of protein. *Colloids Surf. A Physicochem. Eng. Asp.* **2023**, *665*, 131203. [CrossRef]
59. Aasim, M.; Khan, M.H.; Bibi, N.S.; Fernandez-Lahore, M. Understanding the interaction of proteins to ion exchange chromatographic supports: A surface energetics approach. *Biotechnol. Prog.* **2022**, *38*. [CrossRef]
60. Samari-Kermani, M.; Jafari, S.; Rahnama, M.; Raoof, A. Ionic strength and zeta potential effects on colloid transport and retention processes. *Colloid Interface Sci. Commun.* **2021**, *42*, 100389. [CrossRef]
61. Aasim, M.; Bibi, N.S.; Vennapusa, R.R.; Fernandez-Lahore, M. Extended DLVO calculations expose the role of the structural nature of the adsorbent beads during chromatography. *J. Sep. Sci.* **2012**, *35*, 1068–1078. [CrossRef] [PubMed]
62. Persico, M.; Mikhaylin, S.; Doyen, A.; Firdaous, L.; Hammami, R.; Chevalier, M.; Flahaut, C.; Dhulster, P.; Bazinet, L. Formation of peptide layers and adsorption mechanisms on a negatively charged cation-exchange membrane. *J. Colloid Interface Sci.* **2017**, *508*, 488–499. [CrossRef]

63. Firkowska-Boden, I.; Zhang, X.; Jandt, K.D. Controlling Protein Adsorption through Nanostructured Polymeric Surfaces. *Adv. Healthc. Mater.* **2018**, *7*, 1700995. [CrossRef]
64. Jachimska, B.; Pajor, A. Physico-chemical characterization of bovine serum albumin in solution and as deposited on surfaces. *Bioelectrochemistry* **2012**, *87*, 138–146. [CrossRef]
65. Bhakta, S.A.; Evans, E.; Benavidez, T.E.; Garcia, C.D. Protein adsorption onto nanomaterials for the development of biosensors and analytical devices: A review. *Anal. Chim. Acta* **2015**, *872*, 7–25. [CrossRef]
66. Zhao, W.; Poncet-Legrand, C.; Staunton, S.; Quiquampoix, H. pH-Dependent Changes in Structural Stabilities of Bt Cry1Ac Toxin and Contrasting Model Proteins following Adsorption on Montmorillonite. *Environ. Sci. Technol.* **2023**, *57*, 5693–5702. [CrossRef]
67. Zhou, H.-X.; Pang, X. Electrostatic Interactions in Protein Structure, Folding, Binding, and Condensation. *Chem. Rev.* **2018**, *118*, 1691–1741. [CrossRef]
68. Ye, H.; Huang, L.; Li, W.; Zhang, Y.; Zhao, L.; Xin, Q.; Wang, S.; Lin, L.; Ding, X. Protein adsorption and desorption behavior of a pH-responsive membrane based on ethylene vinyl alcohol copolymer. *RSC Adv.* **2017**, *7*, 21398–21405. [CrossRef]
69. Lan, H.; Liu, H.; Ye, Y.; Yin, Z. The Role of Surface Properties on Protein Aggregation Behavior in Aqueous Solution of Different pH Values. *AAPS PharmSciTech* **2020**, *21*. [CrossRef]
70. Li, R.; Wu, Z.; Wangb, Y.; Ding, L.; Wang, Y. Role of pH-induced structural change in protein aggregation in foam fractionation of bovine serum albumin. *Biotechnol. Rep.* **2016**, *9*, 46–52. [CrossRef]
71. Hassel, K.J.; Moresoli, C. Role of pH and Ionic strength on weak cation exchange macroporous Hydrogel membranes and IgG capture. *J. Membr. Sci.* **2016**, *498*, 158–166. [CrossRef]
72. Sushko, M.L.; Shluger, A.L. DLVO theory for like-charged polyelectrolyte and surface interactions. *Mater. Sci. Eng. C* **2007**, *27*, 1090–1095. [CrossRef]
73. Besley, E. Recent Developments in the Methods and Applications of Electrostatic Theory. *Acc. Chem. Res.* **2023**, *56*, 2267–2277. [CrossRef] [PubMed]
74. Luu, H.T.; Perrot, C.; Panneton, R. Influence of Porosity, Fiber Radius and Fiber Orientation on the Transport and Acoustic Properties of Random Fiber Structures. *Acta Acust. United Acust.* **2017**, *103*, 1050–1063. [CrossRef]
75. Blanco, M.A. Computational models for studying physical instabilities in high concentration biotherapeutic formulations. *mAbs* **2022**, *14*. [CrossRef] [PubMed]
76. Quan, X.; Liu, J.; Zhou, J. Multiscale modeling and simulations of protein adsorption: Progresses and perspectives. *Curr. Opin. Colloid Interface Sci.* **2019**, *41*, 74–85. [CrossRef]
77. Liang, J.; Fieg, G.; Jakobtorweihen, S. Molecular Dynamics Simulations of a Binary Protein Mixture Adsorption onto Ion-Exchange Adsorbent. *Ind. Eng. Chem. Res.* **2015**, *54*, 2794–2802. [CrossRef]
78. Taylor, P.A.; Jayaraman, A. Molecular Modeling and Simulations of Peptide–Polymer Conjugates. *Annu. Rev. Chem. Biomol. Eng.* **2020**, *11*, 257–276. [CrossRef]
79. Pink, D.A.; Razul, M.S.G. Computer simulation techniques for food science and engineering: Simulating atomic scale and coarse-grained models. *Food Struct.* **2014**, *1*, 71–90. [CrossRef]
80. Xu, Z.; He, Z.; Quan, X.; Sun, D.; Miao, Z.; Yu, H.; Yang, S.; Chen, Z.; Zeng, J.; Zhou, J. Molecular simulations of charged complex fluids: A review. *Chin. J. Chem. Eng.* **2021**, *31*, 206–226. [CrossRef]
81. Glotzer, S.C.; Paul, W. Molecular and Mesoscale Simulation Methods for Polymer Materials. *Annu. Rev. Mater. Res.* **2002**, *32*, 401–436. [CrossRef]
82. Mills, Z.G.; Mao, W.; Alexeev, A. Mesoscale modeling: Solving complex flows in biology and biotechnology. *Trends Biotechnol.* **2013**, *31*, 426–434. [CrossRef]
83. Khan, M.A.; Sundén, B.; Yuan, J. Analysis of multi-phase transport phenomena with catalyst reactions in polymer electrolyte membrane fuel cells—A review. *J. Power Sources* **2011**, *196*, 7899–7916. [CrossRef]
84. Dalkas, G.; Euston, S.R. Molecular simulation of protein adsorption and conformation at gas-liquid, liquid–liquid and solid–liquid interfaces. *Curr. Opin. Colloid Interface Sci.* **2019**, *41*, 1–10. [CrossRef]
85. Tournois, M.; Mathé, S.; André, I.; Esque, J.; Fernández, M.A. Surface charge distribution: A key parameter for understanding protein behavior in chromatographic processes. *J. Chromatogr. A* **2021**, *1648*, 462151. [CrossRef]
86. Kaiblinger, N.; Hahn, R.; Carta, G. Efficient calculation of the equilibrium composition in multicomponent batch adsorption with the steric mass action model. *Adsorption* **2025**, *31*. [CrossRef]
87. Gama, M.d.S.; Santos, M.S.; Lima, E.R.d.A.; Tavares, F.W.; Barreto, A.G.B. A modified Poisson-Boltzmann equation applied to protein adsorption. *J. Chromatogr. A* **2018**, *1531*, 74–82. [CrossRef] [PubMed]
88. Hahn, R.; Lingg, N. Chapter 21—Separation of proteins by ion-exchange chromatography. In *Ion-Exchange Chromatography and Related Techniques*; Nesterenko, P.N., Poole, C.F., Sun, Y., Eds.; Elsevier: Amsterdam, The Netherlands, 2024; pp. 435–460.

 https://doi.org/10.3390/membranes16010005

89. Briskot, T.; Hahn, T.; Huuk, T.; Hubbuch, J. Protein adsorption on ion exchange adsorbers: A comparison of a stoichiometric and non-stoichiometric modeling approach. *J. Chromatogr. A* **2021**, *1653*, 462397. [CrossRef]
90. Babi, D.K.; Griesbach, J.; Hunt, S.; Insaidoo, F.; Roush, D.; Todd, R.; Staby, A.; Welsh, J.; Wittkopp, F. Opportunities and challenges for model utilization in the biopharmaceutical industry: Current versus future state. *Curr. Opin. Chem. Eng.* **2022**, *36*, 100813. [CrossRef]
91. Boztepe, I.; Yang, X.; Zhao, S.; Kong, L. Module Design for Membrane Chromatography Applications in Downstream Bioprocessing: Flow Distribution Matters. *Ind. Eng. Chem. Res.* **2025**, *64*, 21733–21746. [CrossRef]
92. Hada, K.; Shirzadi, M.; Fukasawa, T.; Fukui, K.; Ishigami, T. Prediction of fluid-particle dynamics and performance in fibrous filters obtained from X-ray CT using convolutional neural network and discrete phase model. *Chem. Eng. J.* **2025**, *514*, 163243. [CrossRef]
93. Mohammadpour, J.; Salehi, F.; Sheikholeslami, M.; Masoudi, M.; Lee, A. Optimization of nanofluid heat transfer in a microchannel heat sink with multiple synthetic jets based on CFD-DPM and MLA. *Int. J. Therm. Sci.* **2021**, *167*, 107008. [CrossRef]
94. Liu, S.; Shikhov, I.; Arns, C. Mechanisms of Pore-Clogging Using a High-Resolution CFD-DEM Colloid Transport Model. *Transp. Porous Media* **2024**, *151*, 831–851. [CrossRef]
95. Yeom, S.B.; Ha, E.-S.; Kim, M.-S.; Jeong, S.H.; Hwang, S.-J.; Choi, D.H. Application of the Discrete Element Method for Manufacturing Process Simulation in the Pharmaceutical Industry. *Pharmaceutics* **2019**, *11*, 414. [CrossRef] [PubMed]
96. Sundaresan, S.; Ozel, A.; Kolehmainen, J. Toward Constitutive Models for Momentum, Species, and Energy Transport in Gas–Particle Flows. *Annu. Rev. Chem. Biomol. Eng.* **2018**, *9*, 61–81. [CrossRef]
97. Tamrakar, A.; Ramachandran, R. CFD–DEM–PBM coupled model development and validation of a 3D top-spray fluidized bed wet granulation process. *Comput. Chem. Eng.* **2019**, *125*, 249–270. [CrossRef]
98. Wodołażski, A. Metaheurystic optimization of CFD–multiphase population balance and biokinetics aeration stirrer tank bioreactor of sludge flocs for scale-up study with bio(de/re)flocculation. *Biochem. Eng. J.* **2022**, *184*, 108477. [CrossRef]
99. Zeng, L.; Franks, G.V.; Goudeli, E. Aggregation and breakage dynamics of alumina particles under shear by coupled Computational Fluid Dynamics—Discrete Element Method. *J. Colloid Interface Sci.* **2024**, *661*, 750–760. [CrossRef] [PubMed]
100. Zhang, L.a.; Diao, Y.; Jiang, J.; Chu, M.; Han, K.; Shen, H. Study on turbulent aggregation dynamics in the process of fine particles capture by single fibers. *Part. Sci. Technol.* **2022**, 1–9. [CrossRef]
101. Kumar, A.; Gautam, S.; Atri, S.; Tafreshi, H.V.; Pourdeyhimi, B. The impact of particle deposition on collection efficiency of electret fibers. *J. Aerosol Sci* **2024**, *181*, 106426. [CrossRef]
102. Briskot, T.; Hahn, T.; Huuk, T.; Hubbuch, J. Adsorption of colloidal proteins in ion-exchange chromatography under consideration of charge regulation. *J. Chromatogr. A* **2020**, *1611*, 460608. [CrossRef]
103. Lohaus, J.; Perez, Y.M.; Wessling, M. What are the microscopic events of colloidal membrane fouling? *J. Membr. Sci.* **2018**, *553*, 90–98. [CrossRef]
104. Agmo Hernández, V. An overview of surface forces and the DLVO theory. *ChemTexts* **2023**, *9*. [CrossRef]
105. Mahian, O.; Kolsi, L.; Amani, M.; Estellé, P.; Ahmadi, G.; Kleinstreuer, C.; Marshall, J.S.; Siavashi, M.; Taylor, R.A.; Niazmand, H.; et al. Recent advances in modeling and simulation of nanofluid flows-Part I: Fundamentals and theory. *Phys. Rep.* **2019**, *790*, 1–48. [CrossRef]
106. Pellicane, G. Colloidal Model of Lysozyme Aqueous Solutions: A Computer Simulation and Theoretical Study. *J. Phys. Chem. B* **2012**, *116*, 2114–2120. [CrossRef]
107. Sadegh, H.; Sahay, R.; Soni, S. Protein–polymer interaction: Transfer loading at interfacial region of PES-based membrane and BSA. *J. Appl. Polym. Sci.* **2019**, *136*, 47931. [CrossRef]
108. Aasim, M.; Hidayatullah Khan, M.; Bibi, N.S.; Zaman Khan, N. Protein adsorption onto monoliths: A surface energetics study. *Eng. Life Sci.* **2018**, *18*, 179–186. [CrossRef]
109. Tsapikouni, T.S.; Missirlis, Y.F. Protein–material interactions: From micro-to-nano scale. *Mater. Sci. Eng. B* **2008**, *152*, 2–7. [CrossRef]
110. Song, Y.; Li, A.; Cui, H.; Zhou, B.; Qiao, J.; Wei, J.; Li, X. Protein language model empowered the robust ASR-driven PET hydrolase featured with two PET binding motifs. *Green Carbon* **2025**. [CrossRef]
111. Bradford, S.A.; Sasidharan, S.; Kim, H.; Gomez-Flores, A.; Li, T.; Shen, C. Colloid Interaction Energies for Surfaces with Steric Effects and Incompressible and/or Compressible Roughness. *Langmuir* **2021**, *37*, 1501–1510. [CrossRef]
112. Misiura, A.; Dutta, C.; Leung, W.; Zepeda O, J.; Terlier, T.; Landes, C.F. The competing influence of surface roughness, hydrophobicity, and electrostatics on protein dynamics on a self-assembled monolayer. *J. Chem. Phys.* **2022**, *156*. [CrossRef]
113. Mollon, G.; Quacquarelli, A.; Andò, E.; Viggiani, G. Can friction replace roughness in the numerical simulation of granular materials? *Granul. Matter* **2020**, *22*, 42. [CrossRef]

114. Akampumuza, O.; Xu, H.; Xiong, J.; Zhang, H.; Quan, Z.; Qin, X. Analyzing the effect of nanofiber orientation on membrane filtration properties with the progressive increase in its thickness: A numerical and experimental approach. *Text. Res. J.* **2020**, *90*, 24–36. [CrossRef]
115. Mukherjee, R.; Sansare, S.; Nagarajan, V.; Chaudhuri, B. Discrete Element Modeling (DEM) based investigation of tribocharging in the pharmaceutical powders during hopper discharge. *Int. J. Pharm.* **2021**, *596*, 120284. [CrossRef] [PubMed]
116. Oyegbile, B.; Ay, P.; Narra, S. Flocculation kinetics and hydrodynamic interactions in natural and engineered flow systems: A review. *Environ. Eng. Res.* **2016**, *21*, 1–14. [CrossRef]
117. McKenzie, J.L.; Webster, T.J. *Protein Interactions at Material Surfaces*; Springer: New York, NY, USA, 2009; pp. 215–237.
118. Barberi, J.; Spriano, S. Titanium and Protein Adsorption: An Overview of Mechanisms and Effects of Surface Features. *Materials* **2021**, *14*, 1590. [CrossRef]
119. Roberts, D.; Keeling, R.; Tracka, M.; Van Der Walle, C.F.; Uddin, S.; Warwicker, J.; Curtis, R. The Role of Electrostatics in Protein–Protein Interactions of a Monoclonal Antibody. *Mol. Pharm.* **2014**, *11*, 2475–2489. [CrossRef]
120. Singh, P.; Roche, A.; van der Walle, C.F.; Uddin, S.; Du, J.; Warwicker, J.; Pluen, A.; Curtis, R. Determination of Protein–Protein Interactions in a Mixture of Two Monoclonal Antibodies. *Mol. Pharm.* **2019**, *16*, 4775–4786. [CrossRef]
121. Bansal, R.; Srivastava, P.; Rathore, A.S.; Chokshi, P. Population balance modelling of aggregation of monoclonal antibody based therapeutic proteins. *Chem. Eng. Sci.* **2020**, *216*, 115479. [CrossRef]
122. Ren, J.; Way, A.; Wang, C. Enhancing aggregate reduction using anion exchange hybrid filter in an immunocytokine diabody fusion protein purification process. *J. Chromatogr. A* **2025**, *1739*, 465526. [CrossRef]
123. Vlachy, V.; Kalyuzhnyi, Y.V.; Hribar-Lee, B.; Dill, K.A. Protein Association in Solution: Statistical Mechanical Modeling. *Biomolecules* **2023**, *13*, 1703. [CrossRef] [PubMed]
124. Pusara, S.; Yamin, P.; Wenzel, W.; Krstić, M.; Kozlowska, M. A coarse-grained xDLVO model for colloidal protein–protein interactions. *Phys. Chem. Chem. Phys.* **2021**, *23*, 12780–12794. [CrossRef]
125. Sharma, I.; Pattanayek, S.K. Effect of surface energy of solid surfaces on the micro- and macroscopic properties of adsorbed BSA and lysozyme. *Biophys. Chem.* **2017**, *226*, 14–22. [CrossRef] [PubMed]
126. Subeshan, B.; Atayo, A.; Asmatulu, E. Machine learning applications for electrospun nanofibers: A review. *J. Mater. Sci.* **2024**, *59*, 14095–14140. [CrossRef]
127. Dangayach, R.; Jeong, N.; Demirel, E.; Uzal, N.; Fung, V.; Chen, Y. Machine Learning-Aided Inverse Design and Discovery of Novel Polymeric Materials for Membrane Separation. *Environ. Sci. Technol.* **2025**, *59*, 993–1012. [CrossRef]

MDPI AG
Grosspeteranlage 5
4052 Basel
Switzerland
Tel.: +41 61 683 77 34

Membranes Editorial Office
E-mail: membranes@mdpi.com
www.mdpi.com/journal/membranes

www.ingramcontent.com/pod-product-compliance
Lightning Source LLC
Chambersburg PA
CBHW062054220926
51850CB00010B/2762